Fermi's Paradox is Bullshit

the Evidence for Extraterrestrial Life

Ken Goudsward

ISBN: 978-1-989940-54-9
© 2022 Ken Goudsward
Dimensionfold Publishing
dimensionfold.com

Cover art based on a Photo by
Bruce Warrington on Unsplash

Acknowledgments

I would like to thank the many people with whom I have discussed ufology and astronomy over the past couple of years, particularly Brad Turner, Jeremy Stewart, Dana DeGans, Mark Eddy, Scott Stewart, Barbara DeLong, Robert Kalil, Rev. Michael J.S. Carter, and Jim Willis.

I would also like to thank Salem Webb for his assistance during the developing the chapter on logical fallacies.

Introduction

There is a 'thing' called the "Fermi Paradox", or "Fermi's Paradox". I call it a 'thing' because it isn't really a paradox, and it isn't really a theory, and it isn't exactly an idea. What exactly is it? I suppose the *closest* of these terms is that it is an idea. The idea is that there is a logical flow of thought that 'proves' that we should doubt the existence of extraterrestrial life. The idea is based on several problematic sub-ideas. To examine these problems is the main purpose of this book.

First we need to outline what the so-called idea says. As defined on Wikipedia: "*The Fermi paradox is the discrepancy between the lack of conclusive evidence of advanced extraterrestrial life and the apparently high a priori likelihood of its existence, and by extension of obtaining such evidence*". Essentially, this is seven separate statements rolled into one.

1. There is a strong probability of the existence of advanced extraterrestrial life.
2. There is a high level of confidence in our ability to obtain evidence of the existence of advanced extraterrestrial life.
3. There is a lack of conclusive evidence of advanced extraterrestrial life.
4. This supposed lack of evidence appears to disagree with statement #2.
5. This seems strange for some reason.
6. It must be a paradox.
7. Therefore, there probably isn't any advanced extraterrestrial life after all.

1

By the end of this book, we will be able to address each of these statements within their combined framework of this so-called paradox. Along the way we will examine some of them directly or indirectly, by examining all the available evidence. To begin with, let's get statement number six out of the way. "Fermi's paradox" is not a paradox, nor was it proposed by Fermi. This is the only thing paradoxical about it.

A paradox is defined by the Merriam-Webster Dictionary as follows:

1. a statement that is seemingly contradictory or opposed to common sense and yet is perhaps true.

2. a self-contradictory statement that at first seems true.

Ironically, and paradoxically, the dictionary seems to be giving us a paradoxical definition. Meaning 1 is a true statement that seems false. And meaning 2 is a false statement that seems true. I suppose if we combined the two definitions we could get a statement whose truth can not be determined. Wikipedia says "A paradox is a logically self-contradictory statement or a statement that runs contrary to one's expectation." By this definition we can include statements which are difficult to determine the truth of, and frankly, any statement we find surprising. This is starting to get nonsensically inclusive. So maybe the paradox is a paradox? Or maybe it just makes no sense? Or maybe it surprised somebody one time? Can we at least get a definition of one word?

Mathematics is a field with very strict definitions. In the mathematical study of logic, a paradox is more strictly defined. Mathematicians may use paradoxes to show the use of invalid arguments as a learning tool to promote critical thinking, or to reveal errors in definitions that were assumed to be rigorous, and therefore to cause axioms of mathematics and logic to be re-examined. By this definition, the paradox is a fundamentally false statement. By this definition, perhaps "Fermi's Paradox" is truly a paradox. It is a fundamentally wrong statement based on faulty logic and/or faulty axiomatic assumptions. The trouble is that the definition of 'paradox' has been allowed to include the opposite meaning; weird but true facts that seem surprising. This has come to be the primary usage among non-mathematicians. Sorry, but it's wrong, and that is also not a paradox.

Italian-American physicist Enrico Fermi's name is associated with 'the paradox' because of a casual conversation he had in the summer of 1950 with a few of his colleagues. While on a lunch break, they had been discussing recent reports of flying saucers and whether faster-than-light travel might be theoretically possible. Fermi is reported to have said something to the effect of "But where is everybody?" (the exact quote is uncertain). Fermi did nothing more than ask a question. A mere question is and has never been a paradox. It is unclear to me whether Fermi believed in extraterrestrial life or not. He did not, to the best of my knowledge, author any books or papers on the subject of UFOs or extraterrestrial life. He was not a ufologist, but rather a nuclear physicist, and one one of the world's top experts at that.

Since the incident at Roswell New Mexico in 1947, and in the three years ensuing until Fermi posed his notorious question, a number of additional sightings allegedly occurred and were widely publicized, including the Kenneth Arnold sighting, the Mantel incident, and several other notable cases[1], as well as hundreds of less publicized reports. The FBI had gotten involved in keeping files of these reports, but according to *the Guardian*, by 1949 "The FBI was so overwhelmed with sightings of flying saucers in the 1940s that agents routinely destroyed reports because of lack of filing space." Suffice to say, the idea of UFOs was very much a mainstream topic in the late forties and early fifties. It is not in any way remarkable that any scientist or citizen would be talking about it, and with a frustrating lack of hard evidence (whether due to conspiracies of coverup or not), such conversation necessarily involved a lot of wondering, questioning, and speculation.

I do not ascribe to Fermi any sort of assumptions on the matter, nor any personal agenda to prove a set of beliefs either one way or the other. I believe Fermi to have been a curious and logical man of science, and I think his career speaks for itself as proof of this. However, and Fermi's own writing does attest to this[2], he appeared to be all too cognizant that not everyone was as logical nor as curious as he.

Aliens Must Not Exist.

Nowadays, "the paradox" is generally used to 'prove' that aliens must not exist. It is difficult to ascertain when this assertion began to take hold in America. It was certainly not the case in 1947, but the flurry of government

---

[1] https://en.wikipedia.org/wiki/UFO_sightings_in_the_United_States

[2] Fermi, Enrico (2004). "The Future of Nuclear Physics". In Cronin, J.W (ed.). Fermi Remembered. Chicago: University of Chicago Press. ISBN 978-0-226-12111-6.

4

and media activity with regards to the UFO phenomenon after Roswell quickly had a polarizing effect. Shortly thereafter, the official channels magnetized to the side of denial, even while attempting to maintain an air of neutrality with the existence of such programs as Project Sign and Project Grudge, which would later morph into the now infamous Project Blue Book. Personally, I find it interesting that a "Sign" quickly turned into a "Grudge". How more obvious can you get? But the existence of aliens and their presence on American soil had quickly become a political problem, not a scientific one. America stands on a foundation of power, and that power became threatened by a foreign power so advanced as to be inconceivable.

If the aliens are able to invade our airspace undetected, there is no telling what kind of damage they could inflict if hostile. Furthermore, if the American military were to engage in battle against these strange technologies there would be no contest in terms of who would win. The superior technologies of the aliens were enough to shut down any foolish notions the military may have initially entertained. Interestingly, this is still the case, as evidenced by the popular verbiage in use as of late. At the time of this writing in 2022, the last two years have seen a dramatic shift in governmental attitudes. The UFOs have become UAPs, and various governmental agencies now speak openly in the press about "the UAP threat".

In contrast, after Roswell measures were taken to brush ufology under the rug and to completely discredit anyone who still took the matter seriously. This occurred in many forms and in many segments of society, one of which was academia. The search for intelligent alien life has been underway by the academic community and independent scholars for hundreds of years. As far back as 1896, Nikola Tesla was actively involved in developing his new technologies for such a purpose. One might wonder if this search was part of what drove such astronomers as Galileo and Copernicus. Throughout history, our quest to get a better view of space has consistently attracted the brightest minds of their time. What was everybody looking for? There seems to be an underlying subconscious knowledge that there is "something" out there worth pursuing, and that someday perhaps, if we can just build a better telescope, we might find it.

We Would Be Able To Find Them

One of the main assumptions at the core of this book is encapsulated in the heading above. This high level of confidence in our ability to obtain evidence of the existence of advanced extraterrestrial life plays a key step in the logic of the argument behind "the paradox". We seem to feel that we are capable of anything. No, let me refine that. We seem to think that we are *already capable* of anything and have *already tried* the right thing.

This is the most unscientific way of thinking possible. Science is all about thinking fresh ideas, trying something new, and improving methodology. Science is exploration. Science knows we are limited and we need to keep pushing these limits. We still haven't really figured out how gravity works, but no scientist is going to claim that it doesn't exist. Yet this is exactly what is being done by the so-called Fermi Paradox.

One way of summarizing the 'paradox' is that "since we haven't found any extraterrestrial life, it must not exist", or "if there were aliens, we would have found them by now". Science doesn't work that way. Sadly though, this is one of the most common ways that the 'paradox' seems to be referenced in conversation. Many people think that it proves that aliens don't exist. It doesn't. The rest of this book will concern itself with examining why it doesn't.

Where Is Everybody?

I have very strong doubts that Fermi ever asked such a question. If Fermi actually did ask this question, he seems to have gone out of his way to ignore a great deal of evidence. This is particularly true because of the time he supposedly asked it. I would be more willing to believe such a claim, were it pinned to sometime in the seventies or eighties, by which time a pall of misinformation and ridicule had been established, but as already noted, in 1950 there was little censorship of UFO phenomena. This brings me to what I see as the real paradox involved in this discussion.

It is a paradox that any scientifically minded person could possibly act as though there were no evidence for extraterrestrial intelligence. This is a paradox because science is all about the evidence. Experimental evidence is one hundred percent of the foundations of all science. Let me be clear. There is an abundance of evidence in and around the search for life in space. This evidence includes but is not limited to the UFO phenomena, of which the sheer volume of evidence is overwhelming. I am not saying that all of the evidence is relevant. I am not saying that none of the evidence has been intentionally falsified. I am not saying that all the evidence is clear, and I am not saying that the evidence necessarily points to any one conclusion. Just that there is *a ton* of evidence. This evidence presents itself both in space and here on earth. Thus, one can look at the evidence in two general piles, and divide it into two separate though obviously related questions:

Are they THERE? & Are they HERE?

Are They There?

The fundamental question at hand is whether or not life exists, or indeed has ever existed, anywhere in the universe other than here on earth. At once we must acknowledge that this question raises philosophical and religious flags. Many religions teach that all life on earth was intentionally and specifically created by someone and that this is thought to have occurred only once, here on earth. Thus, we have the special privilege of being the only intelligent beings and there is no one else "out there". I have no wish to attempt to refute this claim itself; only its corollary.

Suppose you have managed to avoid the trap of religion. You may not even be aware that you are most likely culturally bound under the same false corollary. That is to say, the culture of the secular English-speaking world has been strongly influenced by such philosophical movements as the enlightenment, neoclassicism, humanism, socialism, democracy, and even Darwinian evolution to subconsciously reinforce the idea that *man is the pinnacle, not of creation, but of all of nature.*

According to this dogma, we humans are the best we have ever been, and are vastly superior in every way to anything that is non-human. This has been indoctrinated into Western Civilization for thousands of years, extending back at least to the Greeks. This kind of thinking is incredibly damaging to each of us individually, and to society as a whole, not to mention to our planet. It is wildly arrogant and narcissistic, and it blocks our creativity and curiosity.

This high level of arrogance is what I see when I hear the words "where is everybody?" Can we really be so arrogant as to think that we know all there is to know? We have it all figured out. There's no one as smart as us. Whether though religious or humanist bias, we have been led to this false conclusion, and here we stand, proud in our own ignorance.

Are we the smartest beings in the universe? Are we even the smartest beings on earth? Perhaps not. There are many animal species now being recognized as both sentient but sapient. That is, they are aware of their own sentience.[3] Homo Sapiens has been dethroned. We are among peers[4]. You don't have to watch "the news" every night to realize that humans might actually be one of the stupidest species on earth. With all this going on right here on *our* planet, who can say what is happening elsewhere?

Logical Fallacies

Both sides of the ET argument commonly employ a number of logical fallacies. The nice thing about logic is that it is a form of mathematically rigorous language. The unfortunate side of logic is that it is all too often employed in sloppy, non-rigorous ways, and most people lack the training (or the time) to recognize the improper use when they see it. Without getting terribly bogged down in philosophical minutia, I feel it worthwhile to include a brief table of exactly which of the logical fallacies are in play by both sides of the debate. Of the twenty four most common

---

[3] https://en.wikipedia.org/wiki/Animal_consciousness

[4] https://www.gov.uk/government/news/
lobsters-octopus-and-crabs-recognised-as-sentient-beings

logical fallacies listed at https://yourlogicalfallacyis.com/, no less than twelve are employed regularly in the ET debate.

Fallacy	Used by Believers	Used by Deniers
Appeal To Authority	YES	YES
the Texas Sharpshooter	YES	YES
Begging the Question		YES
False Cause		YES
Burden of Proof		YES
the Fallacy Fallacy		YES
Personal Incredulity		YES
Ad Hominem	rarely	often
Anecdotal	YES	
Bandwagon	YES	YES
Appeal To Emotion	YES	YES

The Appeal To Authority

Both sides defer to "experts" using the appeal to authority in order to back their claim when they know of no evidential proof. This is very common, and frankly unavoidable when dealing with any questions of complexity. It is normal and necessary to do this. What I mean is that each human every day relies on preauthorized assumptions. When I wake up in the morning I remember that the sun rises in the east, and that my name is Ken, and a million other miniscule details. I also default to a vague conceptualization of how I have previously settled on some of the "bigger picture" issues and questions. What is my stand on political issues or religious tenets or matters of taste and style? Without having to examine these in detail each morning we rely on some kind of intuition of our own memory. We have thought these issues through at some point and stored a simplified answer somewhere in our mind. We do not need to rehash these questions every day, nor could we even if we wanted to. There aren't enough hours in a day to even scratch the surface. We would become crippled by indecision. The level of complexity of being a living sentient being demands a simplification and reduction of prior thought into a uniquely personal basic worldview. We do not truly understand how our mind and brain are able to accomplish this amazing task, but each of us contains a working model of the universe around us, in a compressed data format.

In forming this model of the universe we consider all kinds of data. Some of this data is essentially what would be properly considered scientific evidence. Not all of it is though. We utilize the opinions of others, especially friends and family and those we look up to. This is a very useful method of gleaning information, particularly complex information. The entire body of human knowledge can be addressed by leveraging the time and energy of those who have spent their lifetimes examining specific fields. We need not reproduce their time and energy. We can simply use their results. I do not need to become an accountant and a lawyer and a doctor and a mechanic. I simply need to *know* an accountant and a lawyer and a doctor and a mechanic. Likewise for chemists, biologists, psychologists, ecologists, economists, war historians, epidemiologists, epistemologists, and estheticians. We all rely on the expertise of others. The critical need here is to realize and admit that we must do it and are doing it. Of course, the only reason Enrico Fermi is linked to this discussion at all is because someone decided to appeal to authority by dropping his name, he being a famous physicist and widely regarded as a very clever fellow.

When it comes to the questions of extraterrestrial life, there is no one discipline that the question falls squarely into. A physicist is not the most qualified person to address this question. A chemist may have some pertinent insight, as may an astronomer. A mathematician may think his answer seems relevant, as may a philosopher or a priest. Our own personal interests and leanings may influence our choice of who we consider to hold relevant opinions on the topic.

the Bandwagon

The term "bandwagon" is used derogatorily in reference to the adoption of an opinion that is currently fashionable or popular and attracting increasing support. This is the realm of "the influencer". The trend of influencers is by no means a new one. There have always been people whose expertise lies in affecting the perceptions of others. The process is similar to the process we just discussed under the auspices of expertise. The difference between the appeal to authority and the bandwagon, is that while influencers have perceived authority they often have little to no actual expertise.

A second differentiation with the bandwagon effect is that a literal singular influencer may not exist. A bandwagon need not be centralized around a single influencing authority figure. Bandwagons often form, seemingly in a vacuum, based entirely on imperceptible subtle currents within culture. Movements may begin in grassroots groundswells that may take years or even decades to reach a point of recognizability. Sometimes there are notable figures who help it along but by no means were the originators of the idea. Sometimes the ideas are hard fought for recognition by thousands of nameless heroes who have devoted their lives to a cause. This has been an ongoing effort on the part of millions of people on both sides of the ET and UFO debates. The foot soldiers in this battle may employ logic, complete with fallacies. They have many other weapons in their arsenals as well, including politics, religious biases, fear, hope, ridicule, and even outright violence. The more subtle influencer may just tell a story.

Anecdotal Evidence

Arguments from anecdotes are generally considered to be a type of logical fallacy, even though this assumption may itself be somewhat tenuous. The supposition here is that anecdotal evidence is a weak form of evidence, and when used as an argument lacking other contributory evidence, is insufficient to determine any conclusive result. The reason for this bias is that the scientific and philosophical community has long valued quantitative methods over qualitative ones. Granted, there is a valid reason for this bias. The scientific method demands repeatability of observation. An experiment must yield similar results regardless of time, place, experimenter, and other extraneous environmental factors before any conclusions can be drawn.

In the absence of a controlled laboratory environment in which to reproduce a given phenomenon, it becomes difficult to frame any conclusions within a scientific framework. However, this is not to say that the phenomenon itself is unscientific in any way, merely that it currently lacks a dependable method of collecting data. This status is actually an indication of a proto-scientific phenomenon, rather than an unscientific one. What I mean by this is that the phenomenon in question does not yet have reliable technological tools for reading it. Imagine you are living a few hundred years ago. There are no thermometers. Your neighbor comes in and tells you it's cold outside. Can you believe him? Maybe you do or maybe you don't, because all you have to go on is his opinion and his description of his experience. He has no way of quantifying the data. The next day, the temperature seems to have dropped even further, and your neighbor comes back and tells you "it's so cold the pond is frozen." Now, you don't have to simply take his word for it — you can go outside and look at the pond yourself. This is the first step toward a measurable and quantifiable tool for measuring temperature. It is a very blunt measurement as it really only can differentiate between "cold enough to freeze water" and "not cold enough to freeze water." These are the types of blunt tools that have been used for nearly the entirety of human history.

In ancient times, weight was important for fair economical trading. A simple scale was used to simply compare the weights of two objects. These scales were essentially just a balance. This is the type of scale you may recall seeing held by "Lady Justice", or used by Anubis in the Egyptian "Book Of The Dead". This type of scale had no quantitative measure. You could not tell how heavy an object was intrinsically, or put any sort of numeric value onto it. You could say "object A is heavier than object B" or vice versa. You could speak in generalities. This proved to be sufficient though, with the introduction of "standard weights". A merchant could then compare the product in question to a known predetermined object which could itself be verified by a government inspector. Such standard objects have been in use for around five thousand years, if not longer, and have only very recently been superseded within the last couple of generations primarily due to the electronic revolution.

The simplest form of this development is that of pattern recognition. Let's say I'm selling coffee beans by the kilogram, and I have a balance scale but I do not have a standard set of weights. If I have one bag of coffee, all I need to do is gather up a few rocks and compare them to the coffee with my scale. When I find one that matches, I know it must weigh one kilogram. Now, using that one rock as a comparator, I can fill the rest of my bags with the correct amount of beans. I have found a pattern. The rock matches the original bag of beans, and all my new bags of beans match the rock.

The point of all of this is to say that in order to make a quantifiable measurement, we rely on tools that have been developed technologically and specifically for that purpose. We now can measure mass with great precision. We also have other new tools for measuring force, luminosity, temperature, sound levels, electrical voltage, radio frequency, and just about any other quality you can imagine. Our qualitative word has become quantified by our abundant toolset.

The physical realm — i.e. the realm of physics, is erroneously thought to be a complete package. We tend to think we have identified all possible physical attributes. We think we can measure every possible attribute. This is not true. There are many phenomena that seem to lay beyond the borders of our explanation. Many uncountable mysteries remain. As Robert Dijkgraaf wrote recently in Quanta Magazine, "Recent advances in cosmology allow us to state, with a fair amount of certainty, that 95 percent of the universe is missing."[5]

If my customer queries me on my coffee packaging standards, I can tell him the story of how I found a rock that was the right weight. Will he believe me? If I ask an astrophysicist about dark matter, he will tell me a story about some strange mathematical results that didn't add up, and how various people came up with some guesses as to what might explain the discrepancies. These are not measurements. These are stories. Should I believe him? I probably should, because a lot of intelligent astrophysicists

---

[5] https://www.quantamagazine.org/
contemplating-the-end-of-physics-20201124/

and mathematicians wrestled with the same problems and came up with similar (but not identical) stories.

Now apply this same logic to anecdotal evidence in general. Anecdotal evidence is not necessarily untrustworthy. There are several points to consider when weighing its worth. One of the primary factors is repeatability. If each astrophysicist was telling a very different story, it would be difficult to put much faith in them. (In fact, this does happen. Some of the current proposed ideas around dark matter are wildly divergent from the majority.) When most of the astrophysicists have put forth ideas which converge on a set of similar attributes, the ideas begin to coalesce into a hypothesis. They are all telling the same story which leads to a shared resource pool of testable ideas.

The same is true of ufology. If witnesses all told wildly varying accounts, we wouldn't have much to go on. Instead, as we will examine later in this book, the details of their encounters tend to line up into a small set of well defined variables. UFO witnesses tend to tell the same story. They appear to be approximately as reliable as astrophysicists. This is not to say that any given anecdote could not be misconstrued, imagined, or made up. UFO aficionados and astrophysicists alike, just like the rest of us, can fall victim to our own prejudices and perceptual biases.

the Appeal To Emotion

The Appeal to Emotion is another fallacy that is heavily influenced by social factors, and is frequently employed, whether intentionally or not, by both camps of the ET debate. A wide range of emotions may be involved, from empathy to loneliness, fear, superiority, arrogance, and anger. These emotions may manifest differently depending on one's personal beliefs, their chosen bandwagon, and even their socio-economic status.

As an example, let us examine how fear can be used on either side of the debate. A UFO enthusiast may use fear to encourage a belief that the aliens are our only hope of rescue. They may point out that humanity is destroying the planet, and we are doomed unless some higher power comes from space to teach us the error of our ways, or to take us away to their planet. An ET skeptic on the other hand might seek to instill fear of the aliens themselves, and of their supposed nefarious behaviors. This may be done in a sarcastic or parodying manner, but because emotions work on a level below sarcasm, the fear is effective even when decoupled from logic.

Fear also plays a large role in communicating (or rather choosing not to communicate) our honest opinions on the matter of extraterrestrials. Many UFO experiences have never been reported because the witness was afraid of ridicule. This fear of ridicule is even worse for academics whose careers and livelihood depend on their reputation. A good many scientists likely are forced to play a publicly skeptical role even though they may actually be much more open minded than they appear. If they were to allow open and honest discussion they could easily find themselves discredited and unemployed.

the Texas Sharpshooter

Also known as "cherry picking", this practice is exemplified by the focussing of attention of a certain hand-chosen small subset of the available data that seems to best support one's argument.

A UFO skeptic may point out a few well-known publicly discredited hoaxes in order to cast shade on the entire body of UFO evidence. Even within a given particular case, there are often extraneous details that are more dubious or spurious than the case in general. As an example, when looking at the Roswell incident[6], the crash itself is more supported evidentially than the parts of the story involving alien autopsies. By focusing on the more dubious aspects of the story, a skeptic can easily make the whole incident appear more doubtful.

In a way, anecdotal evidence is an unintended form of cherry picking, used by people who only really know their particular piece of the story and are not well versed in the entire field and the massive body of evidence.

Begging the Question

Begging the question is a particularly misunderstood fallacy. Many people believe that the phrase refers to a question that is "begging for an answer" or in other words a very compelling question. This is not the case however. Begging the question is an ancient phrase which stems from a poor translation of a Greek phrase originally penned by Aristotle: τὸ ἐξ ἀρχῆς, literally meaning "asking for the initial thing". Begging the question is a circular argument where the conclusion is assumed as the beginning premise. As an example, the statement "Green is the best color because it is the greenest of all colors" claims that the

---

[6] see page 90

color green is the best because it is the greenest – which it presupposes is the best. It is a type of circular reasoning: an argument that requires that the desired conclusion be true. This often occurs in an indirect way such that the fallacy's presence is hidden, or at least not easily apparent.

This fallacy sometimes appears in discussions involving space travel. A statement such as "since faster than light travel is impossible, it is therefore impossible for extraterrestrials to travel to earth" contains several unstated assumptions about alien technology and the origins and timelines of any supposed extraterrestrials. These assumptions are not stated, but the base assumption is that it is impossible for aliens to travel the necessary long distances in space.

Personal Incredulity

When a person finds a topic difficult to understand, or is simply unaware of how it works, they may respond as though the topic in question just isn't true. This is the fallacy of personal incredulity.

The necessary technologies for long distance manned space travel are not yet understood by the current human civilization. Our science fiction writers may well have hinted at some of the possibilities, and we might be able to imagine at least vaguely how some of them might work. However, it is very simple for a skeptic to simply state that they are not possible. This is a mistake. Most of what we do today was not possible a hundred years ago. Skeptics back then were dismissive about the ideas of forward thinking visionaries as well. The fallacy of personal incredulity falls into these gaps of knowledge.

This occurs in the case of both skeptics and believers alike. In the case of believers, incredulity may occur when a skeptic presents a rational argument against extraterrestrial life that the believer does not understand or follow. A fallback position of "well I know they're out there" is just as fallacious as "well I know it's impossible".

False Cause

False cause is a type of fallacy in which a correlation or temporal sequence of events is falsely assumed to contain a causal relationship.

In the most general sense, the Fermi Paradox itself is predicated on just such a false assumption by erroneously linking two supposed facts into a causal relationship. The supposed "facts" are:

a) the concept of the presence of intelligent extraterrestrial life in general

b) the current human attempts under the guise of SETI which focus primarily on radio communications.

It is assumed without evidence or proof, that these two motifs are well defined, well understood and determinable. In fact, neither of them are. There are a number of poorly defined concepts strung together here in an illegitimate way. Even the basic concept of intelligence itself is not well understood, even when limiting discussion to terrestrial species. What exactly is intelligent life? No one really knows. We then take a logical leap, and go straight to a completely unrelated concept involving radios. This seems to imply that only civilizations using radios can be considered intelligent. Radios did not exist before 1890. Therefore, Julius Caesar, Aristotle, and Da Vinci must not have been intelligent. Obviously this argument does not add up. And even if it did, the reality is that the radios upon which (b) depends are a flawed methodology. (This will be discussed in greater depth beginning on page 98)

Both (a) and (b) are flawed subconcepts. Now, inexplicably, they are linked together in what seems to be a causal relationship: *"since (b) failed to find evidence, then (a) must be false."*

It is an argument that *seems* to make sense. The only thing sensible about it is that it follows the correct syntax. It fits the template of "since x then y" which gives it the appearance of a well constructed argument. However, each of the unrelated predicates, (each themselves false) are not correlated to each other in any way except that both have been observed by someone and assumed to be true, then thrown into a relational statement with absolutely no justification. This leads us nicely to the next fallacy.

Burden of Proof

The Burden of Proof is not a fallacy in and of itself. In criminal law, there is a legal principle of "the presumption of innocence", commonly summarized as "innocent until proven guilty". In legal matters, the prosecutor has the "burden of proof" to show beyond a reasonable doubt that the accusation is true and correct. The defender does not have the same responsibility and need not present evidence to the contrary. He is presumed innocent.

In logic, the assumption is similar. Any statement should be backed up by evidence. It is incumbent upon the person making any claim to show that there is some reason for their claim. It is the exact same principle. If you are going to claim something as a fact, you need to prove it. (Prove is too strong of a word here but you need to show some evidence and make a viable case, just as a prosecutor must do.)

The other party has no such onus. The accused merely states "I didn't do it". When hearing an unsubstantiated claim, we can simply say "no it isn't". It's up to the claimant to provide evidence. There is no reason for anyone else to try to refute the claim. Claims are "false until proven true."

However, there are a few contradicting principles which come into play. The first is freedom of speech. You have the right to say anything without providing any evidence. That does not make your statement true.

The other factor which seems to be causing some issues here is based on a misunderstanding of the scientific method. A scientific hypothesis is a supposition or proposed explanation made on the basis of limited evidence as a starting point for further investigation. Further research will hopefully gather more data around the phenomena in question. Ideally this research will examine all aspects and all possible alternative explanations. In reality, this may not be practical. Often, research is designed to attempt to either prove or disprove a hypothesis. That is not necessarily "bad" or "wrong". It's typically just a result of how the human mind works. When presented with a new idea, the scientist will ask "what if" questions. These questions may tend to support the new idea or detract from it. In either case, the new research questions derived will hopefully result in the gathering of more data. This doesn't necessarily mean that the scientist is trying to prove or disprove the original idea.

She is gathering evidence, not proof. And yet, there is a certain truth to the idea of disproving. As Liv Grjebine, Ph.D., so succinctly summarized in an article entitled "Why Doubt Is Essential to Science":

"Science, when properly functioning, questions accepted facts and yields both new knowledge and new questions... the real power of science lies precisely in its drive to question and challenge a hypothesis. Indeed, the scientific approach requires changing our understanding of the natural world whenever new evidence emerges from either experimentation or observation."[7]

This facet of investigation and evidence becomes even more apparent in a culture of debunkers and skeptics, necessitated by the prevalence of totally unsubstantiated claims. Thus, it has become acceptable to see the burden of proof resting on the shoulders of the debunkers. Anyone can say anything they like and it is someone else's responsibility to prove them wrong.

This is exactly what has occurred in the case of "Fermi's Paradox". Someone (not Fermi) anonymously put forward a claim without any evidence. Millions have fallen for it unquestioningly and accepted it as science.

Ad Hominem

Ad hominem attacks can take the form of overtly attacking somebody, or more subtly casting doubt on their character or personal attributes as a way to discredit their

---

[7] https://www.scientificamerican.com/ article/why-doubt-is-essential-to-science/

argument. The tactic of an ad hominem attack is to undermine someone's case without actually having to engage with it. This may take the form of bullying, derision, mocking, and name-calling. This is an unethical, unprofessional, and frankly immature form of verbal abuse akin to defamation, slander, and libel.

Sadly, many representatives of academic institutions, government, and the press have deemed it necessary to resort to such low blows. I'm not going to name any names here (and I do see the irony of the burden of proof in my claim), but there has been a long history of official sources using derisive ad hominem attacks against UFO witnesses.

If anything, this tactic really only shows that the attacker lacks any solid evidence or real argument.

The Fallacy Fallacy

Even if a certain person presents a terribly flawed argument, this fact alone does not indicate that their claim is incorrect. As much as I would like to think that anyone mean enough to use the ad hominem argument has to be inherently wrong, to do so would be to fall into the fallacy fallacy. It is incorrect to assume that because a claim has been poorly argued, or a fallacy has been made, that the claim itself must be wrong. It is entirely possible to make a claim that is true and attempt to justify it with various fallacies and poor arguments, just as it is possible to make a claim that is false yet argue with logical coherency for that claim. The truth of the claim itself does not actually depend on the argument presented. To think otherwise is a false cause fallacy.

Summary Of Fallacies

Everyone uses fallacies. Most people do so unintentionally. Some of the more despicable tactics are sometimes used more intentionally by self-proclaimed debunkers, to rather unscientific effect. Ultimately though, the entire debate is heavily flawed on both sides, as is the very notion of Fermi's Paradox itself. We have already discussed this, but here is an opportunity to include a more technical expert opinion. According to Robert A. Freitas Jr. of the Xenology Research Institute in Sacramento, the entire "paradox" itself does not fit the mathematically strenuous definition of a paradox, for two reasons:

The "Fermi Paradox," an argument that extraterrestrial intelligence cannot exist because it has not yet been observed, is a logical fallacy. This "paradox" is a formally invalid inference, both because it requires modal operators lying outside the first-order propositional calculus and because it is unsupported by the observational record.[8]

The first half of Freitas' statement is that there is no paradox. We have already noted that Fermi himself did not propose any such paradox. The only thing paradoxical about Fermi's Paradox is its name. The second point that Fretas makes here is that the claim "is unsupported by the observational record." What exactly is this observational record? That will be the focus of the remainder of this book.

---

[8] Icarus; Volume 62, Issue 3, June 1985, Pages 518-520

Statistical Approaches

We will now examine two opposing arguments that appear to the layman to be based on statistics.

- The Infinity Argument
- The Goldilocks Argument

The other side of the "Where Is Everybody?" coin is the claim that statistically speaking, there *has to* be some intelligent life out there somewhere. I will call this "the Infinity Argument". It goes like this:

a) There are *billions of stars* in the Milky Way similar to the Sun.

b) With high probability, some of these stars have *Earth-like planet*s in a circumstellar habitable zone.

c) Many of these stars, and hence their planets, are much *older* than the Sun. If the Earth is typical[9], some may have developed intelligent life long ago.

d) Some of these civilizations may have developed interstellar travel, a step humans are investigating now.

e) Even at the slow pace of currently envisioned interstellar travel, the Milky Way galaxy could be completely traversed in a few million years.

f) Since many of the stars similar to the Sun are billions of years older, Earth should have already been visited by extraterrestrial civilizations, or at least their probes.

---

[9] and according to point b, it is not atypical

Strictly speaking, this chain of reasoning is entirely feasible. It starts with a factual statement which has been proven to be true. It then adds several layers of evidence-based conclusions, which though perhaps not proven conclusively, are themselves sound claims based on the evidence available to date. This takes us all the way to point (f). There is a huge logical leap here from "it should be possible for aliens to get here" to "they must have been here". This claim is not backed by evidence, is itself a non-scientific claim, and in fact, falls into a "false cause" fallacy, which shall be discussed momentarily.

Due to the formal structure of logic, and the transitive nature of the layered claims, the failure of this final step thus causes the entire argument to fail. It is somewhat amusing to me that whether one attempts to try to prove the existence of extraterrestrial life or to prove their non-existence, both sides employ this same argument. The ET believers will stop at stage (f). "See! They must have been here!". The same argument is employed by the ET deniers. They simply add one final claim, point (g), that goes like this: "If they did exist, they would surely have been here, so since they aren't here, they must not exist!"

Both sides rely on point (f) which we have shown to be invalid. The deniers make two additional mistakes. Point (g) contains a false assumption, and is a perfect example of what is properly known in philosophical circles as "begging the question". This is a very commonly misunderstood fallacy, and to be honest, it hurts my head to think about it. All I will say about this is that the exact phrase within point (g) that falls into this trap is "since they aren't here". We are witnessing circular logic here. Circular logic can be entertaining to watch, but it is not fun, and rather difficult to attempt to counter logically. For more fun information on this fallacy please refer to wikipedia.[10]

As if that weren't enough, point (g) has another problem. It is an invalid dependency. The second half of the sentence is not logically connected in any way to the first half, despite the inclusion of the word "so" which makes it *sound like* it must be related. "My coat is blue, therefore my shirt is also blue" is the same type of statement. It might be true that both my shirt and coat are actually blue. It might also be true that I am wearing a blue coat over a red shirt. There is no actual logical dependency between my coat and my shirt, despite the sentence's attempt to fool me.

This entire fallacious argument is often dressed up in a mathematical guise, under the name "the Drake Equation". The Drake equation attempts to quantify a statistical approach by plugging a variety of entirely made up numbers into a made up formula.[11] The formula itself has a somewhat logical structure at least, and so when the formula is explained it sounds like a sensible mathematical argument. The problem here is that each of the supposed variables are entirely fanciful, with no evidential basis whatsoever. It is an appeal to authority, using the solid reputation of math itself to bolster the opinion of the argument.

Goldilocks

Everyone loves a good story. In a favorite of western culture, we recall "*Goldilocks And The Three*

---

[10] https://en.wikipedia.org/wiki/Begging_the_question

[11] Anders Sandberg, Eric Drexler, Toby Ord; *Dissolving the Fermi Paradox,* arXiv:1806.02404v1 [physics.pop-ph]

Bears" In which the heroine tries various configurations to find the one that is "just right". Hence comes the name of the second statistical approach which seeks to prove almost the complete opposite of the "Infinity Argument".

The Goldilocks argument states that everything is very complicated, and that for life to exist (or for that matter, even for the universe to exist), a very strict set of criteria must be fulfilled within very tight limits. Some proponents of this argument use it as a basis to argue for the existence of an intelligent designer outside of the universe. They will point to a delicate balance between gravity, thermodynamics, nuclear physics, etc, such that the "laws of nature"[12] must have been intelligently programmed for this balance to occur. Other proponents merely use it to support what they see as a very low probability that life could exist beyond earth. It is primarily the second group with which we concern ourselves here.

When it comes down to it, this argument is about the existence of liquid water. For water to exist in liquid form, the location, whether planet or otherwise, must exist within a certain range of temperatures where water is stable in a liquid state. As stated by NASA:

"A celestial object can only orbit so close (like Mercury) or so far (like Pluto) from its star before water on its surface boils away or freezes. The 'Goldilocks Zone,' or habitable zone, is the range of distance with the right temperatures for water to remain liquid."[13]

---

[12] keep in mind, nature does not operate under actual laws. What we see as laws are merely our noticing of consistent and regular patterns of behavior. They explain "what", not "why".

[13] https://exoplanets.nasa.gov/resources/323/goldilocks-zone/

The reason this is a big deal is that any life *that we are familiar with* relies heavily on water for its environment and internal biological processes. It is inescapable that water is essential for the carbon-based biology that we know and love.

I hesitate to even bring up the topic of other possible biochemical metabolisms that use other substrates instead of carbon. This fascinating topic goes well beyond my areas of expertise, and I would encourage the reader to explore it further at this link.[14] However, it is important to note that the effect of the assumption that life requires water puts a limit onto the assumed requirements for life that may not be at all valid. Nevertheless, the bursting of this constraint is so far merely speculation, and is not required to further the argument of this book's thesis.

Water is by no means the only requirement for life. As we all know, animals require oxygen, and plants require carbon dioxide. Besides these there are dozens (maybe hundreds?) of other factors too complex for discussion by an author with limited biology expertise. I prefer to defer to the experts on this matter, and none is better qualified than the eminent molecular biologist Francis Crick who discovered and literally wrote the book on DNA.

"An honest man, armed with all the knowledge available to us now, could only state that in some sense, the origin of life appears at the moment to be almost a miracle, so many are the conditions which would have had to have been satisfied to get it going."

Although Crick did not believe in god, he used the term 'miracle' to convey the seeming impossibility of life

---

[14] https://en.wikipedia.org/wiki/Hypothetical_types_of_biochemistry

forming on earth. His solution to this conundrum was to conclude that life most likely arrived on earth from elsewhere, having traveled through space to get here. Although the idea may at first appear fanciful, Crick apparently viewed this concept in a matter that can be best summed up in the words of the fictional character Sherlock Holmes, as penned by Sir Arthur Conan Doyle: *"Once you eliminate the impossible, whatever remains, no matter how improbable, must be the truth."*

The Right Stuff

As already noted under the Goldilocks heading, there appear to be certain constraints that must exist in order to support life. Non-carbon-based lifeforms[15] notwithstanding, we shall now examine some of these necessary components

Gimme Some Air

We have come to perceive air as a necessary factor for maintaining life. This is not exactly true. In general, animals require oxygen in order to fuel the exothermic reactions within the mitochondria nestled within each cell and provision for the cellular function.

However, in all animals there exist secondary metabolic pathways that bypass the mitochondria, and do not require oxygen. This process is grossly inefficient and sustainable for only brief periods in some types of organs

---

[15] https://en.wikipedia.org/wiki/Hypothetical_types_of_biochemistry

and therefore not highly recommended, but such a workaround does in fact exist.[16]

Furthermore there are species known on earth that utilize novel metabolic pathways on a full time basis and do not require the availability of free oxygen in their environments. As one example, consider the Methanogen[17] class of microorganism that lives in anaerobic environments such as the undersea hydrothermal vents that occur beneath our oceans. These rely on high temperature conditions provided by volcanic activity to energize the community. But there are further examples even in more familiar territory.

In a study published in 2020 in the Proceedings of the National Academy of Sciences, researchers identified an animal that doesn't use oxygen to breathe at all.[18] The small parasitic worm Henneguya Salminicola, lives within the flesh of Chinook salmon. It burns no oxygen, and in fact possesses no mitochondria in which to burn oxygen even if it wanted to. Instead, the organism appears to somehow obtain its energy directly from the host fish, although the mechanism for this is not yet understood.

It turns out that Oxygen maybe isn't as necessary as we thought. But of course we should have already known this. We just need to ask the plants. Plants do not require oxygen. In fact, plants create oxygen as a byproduct of their respiration, during which they take in carbon dioxide. Unlike the animals, most of whom use mitochondria, the

---

[16] https://www.sciencedirect.com/topics/immunology-and-microbiology/anaerobic-metabolism#:~:text=HYPOXIA

[17] https://en.wikipedia.org/wiki/Methanogen#Metabolism

[18] https://phys.org/news/2020-02-henneguya-salminicola-microscopic-parasite-mitochondrial.html

energy for the metabolism of plants comes directly through the sun and is processed using chlorophyll. That's why plants are green. The color arises due to the structure of chlorophyll.

Many bacteria, and some fungi such as yeasts rely on neither of these methods of metabolism, instead utilizing fermentation to provide energy for their cellular functioning. They do not require air at all, and in some cases will die if exposed to air.

Now that we have seen that our assumptions about the necessity for air may need some adjusting, let's ask another question: Where is there air?

It is commonly believed that air exists only on earth. Technically, this is true, but only because the definition of the word air is literally, "the atmosphere of earth". This is a very misleading notion however. Let us replace the word air with atmosphere, a more generic word that means "the layer(s) of gasses that envelop a planet, and are held in place by the gravity of the planet". We now can ask: Where is there atmosphere?

The answer is surprising. Within our own Solar system, nine planets are known to have atmospheres; Mercury, Venus, Earth, Mars, Jupiter, Saturn, Uranus, Neptune, and Pluto. Additionally, our own moon, and the moons of several other planets have atmospheres. This list includes four moons of Jupiter (Io, Callisto, Europa, Ganymede), four moons of Saturn (Titan, Enceladus), one moon of Uranus (Titania), one moon of Neptune (Triton). In addition, one large meteor (Ceres) has an atmosphere.[19]

Not that any of these are eminently breathable to us without a space suit. The moon technically has an atmosphere, but it is "very scant"[20]. Mars' atmosphere is a little thicker, but still only around 1% as thick as earth's.[21] On the plus side, it is primarily composed of carbon dioxide which would make it relatively hospitable to plants, and contains trace amounts of water vapor and oxygen.

Titan has a nice thick and dense atmosphere which could conceivably sustain some kind of life, but not our kind. Other than the nice safe inert nitrogen which makes up over 90 percent of it, it has mostly methane with small amounts of hydrogen, and trace amounts of hydrocarbons, such as ethane, diacetylene, methylacetylene, acetylene, propane, hydrogen cyanide, carbon dioxide, carbon monoxide, many of which are classified as "organic compounds". These chemicals are the basis of all carbon-based life.[22] Might these tiny traces indicate the existence of life far in Titan's past? Some of these same compounds are found in Saturn's atmosphere.

Mercury, Ganymede, and Europa each have very thin atmospheres containing both oxygen and water vapour. In the case of Mercury, the solar wind is constantly pushing the atmosphere away from the planet, so its persistence,

---

[19]

https://en.wikipedia.org/wiki/Atmosphere#Atmospheres_in_the_Solar _System

[20] https://en.wikipedia.org/wiki/Atmosphere_of_the_Moon

[21] but was likely thicker in the past.

[22] https://en.wikipedia.org/wiki/Organic_compound

however fragile, must be maintained by some yet identified source, most likely the planetary crust.

The atmosphere of Callisto is mostly carbon dioxide, but probably also contains oxygen.[23]

The atmosphere of Venus is hot and dense, composed primarily of supercritical carbon dioxide in a fluid, but not quite liquid state.[24],[25]. Could plants adapt to such conditions?

Water, Water, Everywhere

Until quite recently, it was often debated whether there might be water in space, or if earth was unique in sustaining any type of aquatic environment. This was, of course, assumed to be a prerequisite for the development of life anywhere.

The beginning of the end of this debate was instigated in July 4, 2005 when NASA's "Deep Impact" probe slammed into the comet "9P/Tempel 1" at a speed of thirty seven thousand kilometers per hour, kicking up a massive cloud of material that the orbiting parent satellite could sample and photograph. Over the next several years, subsequent data analysis of the imaging readings focused on the chemical content of the debris, and slowly revealed the presence of frozen water[26], carbon dioxide, and organic materials.

---

[23] based on measurements of activity in Callisto's ionosphere

[24] https://en.wikipedia.org/wiki/Atmosphere_of_Venus

[25] https://en.wikipedia.org/wiki/Supercritical_carbon_dioxide

[26] *Investigation of dust and water ice in comet 9P/Tempel 1 from Spitzer observations of the Deep Impact event*
A. Gicquel, D. Bockelée-Morvan , V. V. Zakharov, M. S. Kelley, C. E. Woodward, and D. H. Wooden

These surprising results began to lead many leading scientists to conclude that *"comets may have transported these compounds (water and organic materials) to Earth at one time, playing an essential role in the formation of the solar system and life on Earth."*[27] These conclusions are not based on idle speculation, but on a culmination of several facts regarding the composition of comets, meteors, and asteroids as well as their observed patterns of interaction with Earth and our neighbors. Our moon, and our neighboring planet Mars both are famous for their pockmarked surfaces. Clearly, a lot of stuff has crashed into them over time. Things are no different here on Earth.

According to NASA, it is believed that about 48.5 tons of meteoritic material fall onto Earth every single day. Each of these landings is importing extraterrestrial material. This material is known to include water.

In 2019, comet 2I/Borisov was observed on an earth fly-by. It was spouting off water into space, and doing so at a not insubstantial rate of sixty liters per second. In fact, many comets have been observed to contain water. With all of these watery bodies crashing to earth over millennia, it has been proposed that they may have been the source of around half of Earth's current water supply.

Around 1.8 billion years ago, the massive "Sudbury Comet" crashed into earth, bringing with it a lot of water. Some researchers have estimated the amount as up to twenty-five quadrillion liters of water. That's about two percent of our current water, which may not sound like a lot, but this one impact would have made a substantial and

---

[27] quoted from the NASA Jet Propulsion Lab website - https://www.jpl.nasa.gov/missions/deep-impact

noticeable difference to the earth's sea levels overnight. Considering the 4.6 billion years that have elapsed on earth, it is not surprising that eventually, an almost uncountable number of smaller comet impacts would certainly be adequate to account for the entirety of the earth's hydrosphere.

Another factor to consider is the fact that water could not have existed on the early earth. It had to have been added much later. The inner planets, Mercury, Venus, Earth, and Mars, are made mostly of iron and silicates whose condensation temperatures are not compatible with that of water. The temperatures that must have been present during the planetary formation period simply did not allow for water condensation. Any water present would have been in a gaseous form and would not have congealed along with the solid materials that formed the clumpy protoplanets. It was physically impossible for oceans to form on the early Earth.

Further away from the young Sun, temperatures were cooler, and water could condense and form icy planetesimals. The boundary of the region where ice could form in the early Solar System is known as the frost line. This imaginary line is located between Mars and Jupiter, where the asteroid belt now orbits. Therefore, the earliest water and ice in our solar system formed in this area or further out within the gas giants and their moons. Thus it is not surprising that it is in this area that we are now finding water-rich meteors and comets in this zone. These are the objects that began delivering water to Earth.

Water imports are not only one way that comets and meteors affect earth's water supply and sea levels. The Chicxulub asteroid and the Eltanin impact object both impacted in the ocean, causing massive gassification and evaporation, as well as enormous tsunamis that ejected a lot of water out of the ocean. Longstanding climate changes following these events also affected the atmospheric water load and glaciation.

The water cycle on earth is complex, utilizing solar evaporation, precipitation, glaciation, fluviation, and eventually oceation. All of these factors play a critical role in the shape of the planet and the life therein. Although we can not yet pinpoint the ultimate source of all this water, it is now an accepted fact that there is abundant water in space, and a constant bombardment of extraterrestrial water certainly has played a part in the collection of vast quantities of water here on Earth.

As of 2015, the scientifically confirmed liquid water in our own solar system beyond Earth has been approximated at somewhere between twenty five to fifty times the volume of Earth's water.[28] Much of this water resides on or in the other planets, who have been exposed to the same meteoric water imports.

Mars possesses a readily visible water ice cap in its north polar regions[29], as well as a large ice cap in the southern polar zone which is buried under a cap of frozen carbon dioxide.[30] This discovery was made by the

---

[28] Hall, Shannon (2015). "Our Solar System Is Overflowing with Liquid Water [Graphic]". Scientific American. 314 (6): 14–15. doi:10.1038/scientificamerican0116-14. PMID 27196829.

[29] Carr, M.H. (1996). Water on Mars. New York: Oxford University Press. p. 197.

[30] Bibring, J.-P.; Langevin, Yves; Poulet, François; Gendrin, Aline; Gondet, Brigitte; Berthé, Michel; Soufflot, Alain; Drossart, Pierre; Combes, Michel; Bellucci, Giancarlo; Moroz, Vassili; Mangold, Nicolas; Schmitt, Bernard; Omega Team, the; Erard, S.; Forni, O.;

European Space Agency, after their Mars Express Orbiter mapped the surface of Mars using radar equipment to look for evidence of subsurface water. The readings indicated a water lake about twenty kilometers wide, about a mile under the planet's surface.[31]

Overall, more than five million cubic kilometers of water ice have been detected at or near the surface of Mars. That is enough ice to cover the whole planet to a depth of thirty five meters.[32] In other words, if the polar ice caps and frozen groundwater of Mars were to melt to liquid form, Mars would have an ocean very similar to ours. On top of that, the Carbon Dioxide caps would have had to melt first, thus releasing an atmosphere that could sustain plant life.

Such an ocean likely did in fact exist on Mars a couple billion years ago[33] when its atmosphere was denser

Manaud, N.; Poulleau, G.; Encrenaz, T.; Fouchet, T.; Melchiorri, R.; Altieri, F.; Formisano, V.; Bonello, G.; Fonti, S.; Capaccioni, F.; Cerroni, P.; Coradini, A.; Kottsov, V.; et al. (2004). "Perennial Water Ice Identified in the South Polar Cap of Mars". Nature. 428 (6983): 627–630. Bibcode:2004Natur.428..627B. doi:10.1038/nature02461. PMID 15024393. S2CID 4373206.

[31] Orosei R, Lauro SE, Pettinelli E, Cicchetti A, Coradini M, Cosciotti B, Di Paolo F, Flamini E, Mattei E, Pajola M, Soldovieri F, Cartacci M, Cassenti F, Frigeri A, Giuppi S, Martufi R, Masdea A, Mitri G, Nenna C, Noschese R, Restano M, Seu R (July 25, 2018). "Radar evidence of subglacial liquid water on Mars". Science. 361 (3699): 490–493. arXiv:2004.04587. Bibcode:2018Sci...361..490O. doi:10.1126/science.aar7268. hdl:11573/1148029. PMID 30045881. S2CID 206666385.

[32] Christensen, P. R. (2006). "Water at the Poles and in Permafrost Regions of Mars". Elements. 3 (2): 151–155. doi:10.2113/gselements.2.3.151.

[33] Baker, V. R.; Strom, R. G.; Gulick, V. C.; Kargel, J. S.; Komatsu, G.; Kale, V. S. (1991). "Ancient oceans, ice sheets and the hydrological cycle on Mars". Nature. 352 (6348): 589–594.

and warmer[34], covering around a third of the planet's surface.[35] (Mercury and Venus also likely had water in the past, although it seems unlikely they do currently and the data around the question is rather sparse.)[36]

It is high time we stop thinking of Mars as a dry barren wasteland. All the necessary materials for life are right at hand. Has there ever already been life on Mars? It is certainly possible. Is it possible again? Absolutely, although not without some significant intentionally designed terraforming activity.

Both NASA and the European Space Agency have made it clear that they are already thinking along these lines, though they may not be making public claims about their agenda. Both agencies have ongoing mars exploration programs, with numerous recent missions on Mars including: "Mars Odyssey", "Mars Express", "Mars Exploration Rovers", "Mars Reconnaissance Orbiter", "Mars Phoenix lander", all of which have provided information about water's abundance and distribution on Mars. SpaceX, a private aerospace company under the direction of Elon Musk, is actively engaged in development

Bibcode:1991Natur.352..589B. doi:10.1038/352589a0. S2CID 4321529.

[34] Pollack, J.B.; Kasting, J.F.; Richardson, S.M.; Poliakoff, K. (1987). "The Case for a Wet, Warm Climate on Early Mars". Icarus. 71 (2): 203–224. Bibcode:1987Icar...71..203P. doi:10.1016/0019-1035(87)90147-3. hdl:2060/19870013977. PMID 11539035.

[35] "Ancient ocean may have covered third of Mars". Science Daily. June 14, 2010

[36] https://en.wikipedia.org/wiki/Water_on_terrestrial_planets_of_the_Solar_System

of a manned Mars mission[37] and plans for establishment of a human colony on Mars within the next thirty years. This is not some pipe dream of an uninformed whacko. SpaceX is already subcontracting and delivering on their promises of orbital missions including the manned flights to the International Space Station under contract with NASA. These plans rely on the proven availability of water and carbon dioxide on the "Red Planet".

Mars may be our closest neighboring planet, but it is not the closest solid planetary body capable of harboring water. Of course, our moon has that distinction. Unfortunately, moon exploration went through a very dry spell after NASA's Apollo program completed in 1972, however NASA's new Artemis program intends to send astronauts back to the moon in short order. It is worth noting that currently NASA's only launch vehicles are the ones they hire from SpaceX, so when we say that NASA will be sending men and women to the moon, we really mean that SpaceX will be sending men and women to the moon. This goal is expected to be achieved in 2025. Shortly thereafter, the program will *"establish the first long-term presence on the Moon. Then, we will use what we learn on and around the Moon to take the next giant leap: sending the first astronauts to Mars."*[38]

NASA is explicitly stating the program will logically transform into a manned Mars program. They appear to be on board with Elon Musk's vision. In order for the moon to be a valid proving ground for the Mars missions, the moon missions will need to include several

---

[37] https://www.spacex.com/human-spaceflight/mars/

[38] https://www.nasa.gov/specials/artemis/

critical aspects of the long-term stated goals of Musk. Specifically, SpaceX's reusable rockets will be fueled by refillable fuels that can be manufactured on-planet, that is, right there on Mars and presumably on the moon as well. The source for such fuels is planned to be water. Hydrogen and Oxygen (the component parts of water) can easily be obtained through electrolysis of regular old water.

These will become the fuel for return trips back from Mars. Yes, SpaceX plans to enable two way travel between earth and the Martian and Lunar colonies. This is not intended to be a one way escape from a dying earth, but an first step in a truly cosmic interplanetary civilization of humanity. But does the moon really fit the bill as a proving ground for Mars? Its gravity is similar, its atmosphere is similar, but not as promising as Mars'. What about its water supply?

As early as 1976, there has been scientific evidence to suggest the presence of water on the moon. The Soviet Luna 24 probe gathered subsurface samples, and in February 1978, published their laboratory analysis of these samples showing that they contained 0.1% water by mass.[39] Unfortunately, these results were not acknowledged by the cold-war era western national administrations.

In the decades since, numerous attempts to study the lunar water supply have been made by NASA, the U.S. military, the Japanese, Chinese, and Indian governments, and several academic interests.[40] All of these efforts have

---

[39] Akhmanova, M; Dement'ev, B; Markov, M (February 1978). "Water in the regolith of Mare Crisium (Luna-24)?". Geokhimiya (in Russian) (285).

[40] https://en.wikipedia.org/wiki/Lunar_water

been rewarded with hopeful results that have proved disappointingly and frustratingly difficult to prove conclusively, due in large part to technical details regarding the chemistry of water and similar compounds. The best we can say about most of these observations is "that MIGHT be water". Efforts are ongoing however, with several NASA projects targeting the question in the next year or two.

The most solid results of these tests did actually prove that water vapor is present, at least sometimes, within the very tenuous lunar atmosphere. Unfortunately, the atmosphere doesn't last long on the moon, as it is constantly lost into space. The interesting thing about this is that the water must be leaching out of the ground, in order for it to be there even for a short time.

In 2020, NASA's SOFIA observatory discovered "diffuse water molecules persisting at the Moon's sunlit surface".[41] These molecules are subject to gradual decomposition by sunlight, leaving hydrogen and oxygen lost to outer space.

Before jumping to the outer planets, let us pause again at the asteroid belt, home of the dwarf planetoid, Ceres. Ceres is the largest object in the asteroid belt, orbiting the sun between the orbits of Mars and Jupiter. There is evidence of a significant amount of water beneath the surface of Ceres, as well as in its sparse atmosphere.[42] This atmosphere is generated by a number of areas on the planetoid's surface that have been observed to leak water

---

[41] "NASA - SOFIA discovers water on sunlit surface of the Moon". NASA. 26 October 2020.

[42] https://en.wikipedia.org/wiki/Ceres_(dwarf_planet)#Atmosphere

from beneath the surface at a rate of three liters per second.[43]

Besides Earth, Ceres has the most water of any body in the inner solar system, and likely includes subsurface brine pockets that are potential habitats for life.[44] It is close enough to the Sun to maintain subsurface liquid water for extended periods. Using remote detection methods, Ceres has been found to be rich in carbon, hydrogen, oxygen and nitrogen, which are important elements for the support of organic chemistry.[45]

Both the European Space Agency, and the Chinese Space Agency are designing sample-return missions from Ceres to take place in the next few years, which will undoubtedly provide much informative data. If Ceres can be thought of as a typical example of the myriad asteroids in the belt, the amount of conceivable water contained within the asteroid belt could be staggering.

---

[43] Küppers, M.; O'Rourke, L.; Bockelée-Morvan, D.; Zakharov, V.; Lee, S.; Von Allmen, P.; Carry, B.; Teyssier, D.; Marston, A.; Müller, T.; Crovisier, J.; Barucci, M. A.; Moreno, R. (23 January 2014). "Localized sources of water vapour on the dwarf planet (1) Ceres". Nature. 505 (7484): 525–527. Bibcode:2014Natur.505..525K. doi:10.1038/nature12918. ISSN 0028-0836. PMID 24451541. S2CID 4448395.

[44] Julie C. Castillo-Rogez; et al. (31 January 2020). "Ceres: Astrobiological Target and Possible Ocean World". Astrobiology. 20 (2): 269–291. Bibcode:2020AsBio..20..269C. doi:10.1089/ast.2018.1999. PMID 31904989.

[45] Marchi, S.; Raponi, A.; Prettyman, T. H.; De Sanctis, M. C.; Castillo-Rogez, J.; Raymond, C. A.; Ammannito, E.; Bowling, T.; Ciarniello, M.; Kaplan, H.; Palomba, E.; Russell, C. T.; Vinogradoff, V.; Yamashita, N. (2018). "An aqueously altered carbon-rich Ceres". Nature Astronomy. 3 (2): 140–145. doi:10.1038/s41550-018-0656-0. S2CID 135013590.

Moving past the asteroid belt, Jupiter, the largest planet in our system, also happens to have by far the largest planetary atmosphere in the system. Although it is composed mostly of hydrogen and helium (much like a star) other chemical compounds are present in small amounts, including methane, ammonia, hydrogen sulfide, and water. The water is detectable in small amounts in the high atmosphere, but is thought to increase in the deeper layers which have as yet eluded penetration by instrumentation. These lower atmospheric layers gradually transition into the liquid interior of the planet.

Uranus and Neptune have a lot of water, possibly accounting for over sixty percent of their total mass,[46] — hundreds of times the amount of water found on earth. This water exists in a "supercritical" form, which means that the high temperatures and pressures create a strange fluid that is strictly neither liquid nor gas.

Though barren now, shortly after its formation, Pluto may have had a subsurface ocean, and been able to support life.[47]

Several of the moons in our system also contain water. Europa, one of Jupiter's moons, is thought to have a thick crust of frozen water overlaying a liquid water ocean up to a hundred kilometers deep. These estimates lead to a calculated volume of Europa's oceans of more than twice the volume of Earth's oceans.

---

[46] NASA, On to the Ice Giants. (PDF) Pre-Decadal study summary, presented at the European Geophysical Union, 24 April 2017.

[47] Bierson, Carver; et al. (22 June 2020). "Evidence for a hot start and early ocean formation on Pluto". Nature Geoscience. 769 (7): 468–472. Bibcode:2020NatGe..13..468B. doi:10.1038/s41561-020-0595-0. S2CID 219976751. Retrieved 23 June 2020.

Based on recent observations from Hubble, a subsurface saline ocean is also theorized to exist on Ganymede, another Jovian moon. It is estimated to consist of a 100 km liquid water layer, overlaid with a 150 km thick crust of ice.[48]

One of Saturn's moons, Enceladus, was observed by the Cassini spacecraft in 2005, showing water geysers spouting high into the atmosphere. Further study and analysis confirmed an icy crust and a subsurface ocean.

The bottom line is that there is a lot of water in space beyond earth. According to *Scientific American* magazine, *"It now looks like the solar system is awash with this key ingredient for life. A total volumetric estimate of the confirmed oceans alone adds up to 50 times more water than is found on Earth—a figure that could continue to surge."*[49]

As we have already discussed, comets and asteroids contain water, and since these objects can be found hurtling through space in elliptical orbits, and crashing into planets, it is reasonable to assume that they are a principle transport mechanism for bringing water to planets, including Earth. What has not been clear until now is how the water got into the asteroids and comets in the first place. A study published in Nature last year based on asteroid material sampled during a recent Japanese mission, shows that water is continuously being generated by chemical reactions involving asteroidal oxygen interacting with the passing

---

[48] "NASA's Hubble Observations Suggest Underground Ocean on Jupiter's Largest Moon". NASA. 12 March 2015.

[49] https://www.scientificamerican.com/article/our-solar-system-is-overflowing-with-liquid-water-graphic/

solar wind.[50] The study concluded that Earth's extant water appears to be composed of a mix of this freshly generated 'solar wind water' as well as older cometary core water. This is not the whole picture, however. Another recent study has estimated that, by reason of the same chemical processes, and taking into account the age of the sun, up to half of Earth's water is older than our Solar system.[51] This of course implies that water has come here from outside of our Solar system. Not only is at least some of Earth's water extraterrestrial, it is interstellar. We are not the only star system engaged in a cosmic water trade.

Exoplanets

In July of 2014, NASA confirmed their recent discovery of an atmosphere on a planet officially known as "HD 209458 b"[52]. Evidently, they became tired of this tongue twister rather quickly, as the planet soon was given the nickname "Osiris". It is located around 159 light-years away in the direction of the constellation Pegasus where it orbits a star fairly similar to our sun.

To be clear, the planet itself is not at all similar to earth, being a very hot gas giant that orbits very close to its star. However, it was confirmed as the first extrasolar planet that is known to have an atmosphere. What is this atmosphere like? Well, it contains many elements,

---

[50] Daly, L., Lee, M.R., Hallis, L.J. et al. Solar wind contributions to Earth's oceans. Nat Astron 5, 1275–1285 (2021). https://doi.org/10.1038/s41550-021-01487-w

[51] https://www.scientificamerican.com/article/earth-has-water-older-than-the-sun/

[52] https://en.wikipedia.org/wiki/HD_209458_b

including oxygen and carbon, and is thought to contain water vapor. All the things we thought we needed to support life, then discovered maybe we don't need after all, but are certainly nice to have, are all present on Osiris.

This does not mean that Osiris is necessarily immediately hospitable to human colonization, or that we could walk around without spacesuits. What it does mean is that Osiris does have all the necessary requirements to support life. Not only can Osiris support life in general, but it can support the specific carbon-based organic DNA type of mechanisms that we readily recognize as life. Again, this does not mean that Osiris has life, but it does mean that it could have life. It might *have* life. It might have *had* life. It might *one day* have life.

Osiris is not alone. In 2017, traces of water were discovered[53] within the atmosphere of another exoplanet named 51 Pegasi b, aka. Dimidium. Like Osiris, Dimidium is an extrasolar planet with common characteristics of a new class of planets called hot Jupiters.[54] Also like Osiris, Dimidium is located in the direction of the constellation Pegasus. However, Dimidium is less than a third the distance from us — a mere 50 light-years away.

With our current rocket technology, we could travel to Dimidium, but the journey would take about twenty million years. That's long enough for our rocket bound descendants to evolve considerably. Rockets are not really the way to go though. Asteroids and comets routinely reach

---

[53] Water detected in the atmosphere of hot Jupiter exoplanet 51 Pegasi b". phys.org. February 1, 2017.
[54] Wenz, John (10 October 2019). "Lessons from scorching hot weirdo-planets". Knowable Magazine. Annual Reviews. doi:10.1146/knowable-101019-2. Retrieved 4 April 2022.

speeds up to ten times faster than our rockets. A water bearing comet could travel between Dimidium and Earth in only two million years. That is a span of time that is easily sustainable for a species. To put it into perspective, we have been walking upright for around six million years already. With a colony built on an asteroid, anatomically recognizable humans could make the trip in around seventy-five thousand generations. If they were culturally diligent in their record-keeping, could they conceivably even remember why they were making the journey?

There are many more exoplanets that have been discovered[55], and as of the time of this writing in late 2022, no less than sixty six of these have been analyzed and found to contain water,[56] and fifty seven are deemed "potentially habitable"[57] The closest of these are much closer to us than Dimidium, but we have not yet confirmed the presence of water on them. If water is found on these, the travel time could be cut down by nearly three quarters to six hundred thousand years. In comparison, six hundred thousand years ago our ancestors had begun making crude boats.[58] Could they have possibly imagined such a journey to the stars?

---

[55] There are 5,246 known exoplanets, according to
https://en.wikipedia.org/wiki/List_of_nearest_exoplanets

[56] http://exoplanet.eu/catalog/

[57] https://en.wikipedia.org/wiki
/List_of_potentially_habitable_exoplanets

[58] As evidenced by the fact that Homo Luzonensis would have had to have made a substantial sea crossing over the Huxley Line around 700,000 years ago.

Astrobiology

Since humans first began venturing into space, the search for traces of life has been a topic of particular interest to many scientists and laymen alike. Is there life in space? To some, this is the ultimate question. Oddly though, we may already have the answer.

In 2014, Russian cosmonauts on extravehicular activity maneuvers collected swabs of the exterior hull of the International Space Station, which upon examination, were found to include traces of bacteria and sea plankton. The Russian Space News Agency publicly stated *"these swabs reveal bacteria that were absent during the launch of the ISS module. That is, they have come from outer space and settled along the external surface."*

The Russians did not state how they came to the definitive conclusion that the microbes were not in fact present at launch, which has led to much doubt of the initial claim. In all probability, these bacteria could be terrestrial, even if the launch vessel had been sterilized right before launch, because even the upper atmosphere contains abundant microscopic life.[59] Furthermore the Russians claim to have also found terrestrially originating bacteria "surviving on the space station's external surface, though they had remained within a space vacuum for three years." Although this finding has kind of been brushed past as a minor detail, this really is a big deal. *Bacteria can survive in space.*

This claim has been confirmed by NASA scientist Lynn Rothschild, who stated: *"there is a long history of*

---

[59] https://www.livescience.com/26645-microbes-in-the-sky.html

U.S. and European missions proving that microbes could survive in low Earth orbit for extended periods of time."

Even if these bacteria originated in Earth's upper atmosphere, and were dragged into space by the passing launch vehicle, the implication here is that any object, such as a asteroid, that sweeps through our high atmosphere could theoretically pick up terrestrial microbial life and bring it out into deep space, where it could potentially survive. Furthermore, the same asteroid may eventually crash into another planetary body. We have just established an entirely natural mechanism for life to move from one planet to another.

While this may be technically possible it may be a rather speculative supposition. Is there any evidence of this process actually happening?

Well... actually... Kind of.

Biometeors

Signs of life have been confirmed in comets, meteors, and asteroids many times by many people. though typically met with hearty doses of skepticism and mockery of their colleagues. Richard B. Hoover is a very intelligent and highly acclaimed physicist who has authored no less than thirty-three books and two hundred and fifty papers on subjects as diverse as solar physics, X-ray optics, meteorites, and extremophiles. He holds eleven U.S. patents and has been named Inventor of the Year by NASA, where he worked for forty six years, and is the worlds leading expert on ancient microbial extraction.

Hoover has extracted fossilized microbial microfossils, and chemical biomarkers from frozen Arctic

53

samples, from rare precambrian rocks and from carbonaceous chondrite meteorites. He has managed to discover and extract what he claims are indigenous microfossils of cyanobacteria and other filamentous microorganisms from five distinct meteorites.

As might well have been expected, the establishment and academia have not been quick to substantiate Hoover's findings. His detractors use such ridiculous schoolyard tactics as mudslinging and calling into question his own expertise. Hoover is by all accounts the leader in the world of microfossil extraction.

He has collected, extracted , and cataloged to the satisfaction of all scientific scrutiny, microbial extremophiles and novel bacteria from glaciers and permafrost in Antarctica, Patagonia, Siberia, Alaska and from lakes, geysers and volcanoes in California, Alaska, Crete and Hawaii. He has the auspicious distinction of having discovered, described, and named several new species of bacteria as well as two completely new genera of bacteria and archaea. One of his species had survived undetected by everyone else for thirty two thousand years in a frozen Alaskan pond.

Yet, as soon as meteoric rocks are involved, the pundits dare to claim that Hoover doesn't know what he is doing, and throw his own frontrunning skills back in his face citing a lack of "peer review". What peers? This man is leaps and bounds beyond anyone else in the field! Clearly, it is well worth the effort of listening to Hoover's meteoric claims. Thankfully, there were some within the scientific community who felt that this fascinating topic warranted further investigation. Numerous independent studies have confirmed the presence of organic materials including amino acids within at least one of Hoover's subject meteorites, the Murchison Meteorite,[60] [61] [62] [63]

When it comes to the controversial aspects of meteorite samples, one of the biggest problems is the concern of terrestrial contamination. That is to say, that whenever organic materials are found, there is always the possibility that the sample was contaminated either upon atmospheric entry, terrestrial impact, or in the lab.

This concern was addressed by the Japan Aerospace Exploration Agency when they decided to collect samples directly from asteroids while in deep space. In 2005 the Japanese space probe Hayabusa collected samples from an asteroid named "25143 Itokawa" while in orbit. The

---

[60] Kvenvolden, Keith A.; Lawless, James; Pering, Katherine; Peterson, Etta; Flores, Jose; Ponnamperuma, Cyril; Kaplan, Isaac R.; Moore, Carleton (1970). "Evidence for extraterrestrial amino-acids and hydrocarbons in the Murchison meteorite". Nature. 228 (5275): 923–926. Bibcode:1970Natur.228..923K. doi:10.1038/228923a0. PMID 5482102. S2CID 4147981.

[61] Meierhenrich, Uwe J.; Muñoz Caro, Guillermo M.; Bredehöft, Jan Hendrik; Jessberger, Elmar K.; Thiemann, Wolfram H.-P. (2004). "Identification of diamino acids in the Murchison meteorite". PNAS. 101 (25): 9182–9186. Bibcode:2004PNAS..101.9182M. doi:10.1073/pnas.0403043101. PMC 438950. PMID 15194825.

[62] Engel, Michael H.; Nagy, Bartholomew (29 April 1982). "Distribution and enantiomeric composition of amino acids in the Murchison meteorite". Nature. 296 (5860): 837–840. Bibcode:1982Natur.296..837E. doi:10.1038/296837a0. S2CID 4341990.

[63] Schmitt-Kopplin, Philippe; Gabelica, Zelimir; Gougeon, Régis D.; Fekete, Agnes; Kanawati, Basem; Harir, Mourad; Gebefuegi, Istvan; Eckel, Gerhard; Hertkorn, Norbert (16 February 2010). "High molecular diversity of extraterrestrial organic matter in Murchison meteorite revealed 40 years after its fall" (PDF). PNAS. 107 (7): 2763–2768. Bibcode:2010PNAS..107.2763S. doi:10.1073/pnas.0912157107. PMC 2840304. PMID 20160129.

samples were sealed into a capsule and sent hurtling on a two million kilometer, five year journey back to Earth, where the samples were analyzed in a sterile environment. The sample contained "indigenous organic matter" whose "isotopic compositions indicate a clear extraterrestrial origin."[64] The regolith collected from 25143 Itokawa is thought to have lain there for about eight million years. Given the fact that the orbit of 25143 Itokawa is a near Earth one, this indicates that there has been life in space near earth for at least 8 Million years.

Another meteorite specifically worthy of note is the "Allan Hills 84001" meteorite[65] (hereafter abbreviated ALH84001). This meteorite also was found to contain microscopic fossils of bacteria.[66] That finding has been confirmed in numerous other meteorites, but it is not what makes ALH84001 unique. The interesting thing about this meteorite is that it came from the planet Mars. Chemical analysis suggests that the material of ALH84001 crystallized from molten rock over four billion years ago,[67]

---

[64] Chan, Q; Brunetto, R; Kebukawa, Y; Noguchi, T; Stephant, A; Franchi, I; Zhao, X; Johnson, D; Starkey, N; Anand, M; Russell, S; Schofield, P; Price, M; McDermott, K; Bradley, R; Gilmour, J; Lyon, I; Eithers, P; Lee, M; Sano, Y; Grady, M (2020). First Identification of Indigenous Organic Matter Alongside Water In Itokawa Particle Returned By The Hayabusa Mission. 51st LPSC. Sec. H2O abundance and isotopic composition

[65] https://en.wikipedia.org/wiki/Allan_Hills_84001

[66] Thomas-Keprta, K. L.; Clemett, S. J.; McKay, D. S.; Gibson, E. K.; Wentworth, S. J. (2009). "Origins of magnetite nanocrystals in Martian meteorite ALH84001" (PDF). Geochimica et Cosmochimica Acta. 73 (21): 6631–6677. Bibcode:2009GeCoA..73.6631T. doi:10.1016/j.gca.2009.05.064. Retrieved 2014-05-07.

[67] Lapen, T. J.; et al. (2010). "A Younger Age for ALH84001 and Its Geochemical Link to Shergottite Sources in Mars". Science. 328

when there was liquid water[68] [69] on the planet's Martian surface. So how did it end up here on Earth? The theory is that Mars was impacted by a comet around seventeen million years ago,[70] with sufficient force to knock chunks of rock into orbit. These chunks drifted through space, and eventually ALH84001 fell to Earth about thirteen thousand years ago.

The implications of ALH84001 are four-fold. Firstly, there exist naturally occurring physical mechanisms for the transportation of material between planets. Secondly, there is a known and proven instance of this taking place in our past. Thirdly, Mars used to have an ocean. Fourthly, there was once marine microbial life on Mars.

Organic compounds are hurtling through space all around us, and we have been able to find them almost everywhere we look. NASA's 2006 "Stardust" mission[71] brought back biological samples harvested from a comet's tail, which included many complex biological proteins and amino acids. Similarly, the European Space Agency's Philae lander mission sampled comet 67P in 2014 to reveal

(5976): 347–351. Bibcode:2010Sci...328..347L. doi:10.1126/science.1185395. PMID 20395507. S2CID 17601709.

[68] "The ALH84001 Meteorite". NASA. Jet Propulsion Laboratory. Retrieved 2014-05-07. Orange carbonate grains, 100 to 200 microns across, indicate that the meteorite was once immersed in water.

[69] Eiler, John M.; Fischer, Woodward W.; Halevy, Itay (11 October 2011). "Carbonates in the Martian meteorite Allan Hills 84001 formed at 18 ± 4 °C in a near-surface aqueous environment". Proceedings of the National Academy of Sciences. PNAS. 108 (41): 16895–16899. doi:10.1073/pnas.1109444108. PMC 3193235. PMID 21969543.

[70] "How could ALH84001 get from Mars to Earth?". Lunar and Planetary Institute. LPI. 2014. Retrieved 2014-05-07.

[71] https://www.jpl.nasa.gov/missions/stardust

sixteen organic different compounds.[72] The Leonid meteors have also been sampled and found to contain organic material.

The story gets even stranger though. Organic matter has been found on asteroids and comets, but recent findings by NASA and others suggest that DNA and RNA components such as adenine, guanine and related organic molecules — the building blocks for life as we know it — may be formed within pregalactic nebulae, and even in the vast empty stretches of outer space.[73] [74]

Nebulae of Life

A nebula is a giant cloud of dust and gas in space. Nebulae are regions where new stars are beginning to form and are often referred to as "star nurseries." Nebulae are also left behind as gas and dust thrown out by the explosion of a dying star, such as a supernova. Essentially, nebulae are the afterlife and preformed reincarnation of stars. They are undead stars. This makes all the more remarkable what has recently been discovered in nebulae.

Carl Sagan was quoted as saying "We are made of star-stuff." He referred to the fact that all elements are produced within the nuclear fusion furnaces of stars, so

---

[72] Jordans, Frank (30 July 2015). "Philae probe finds evidence that comets can be cosmic labs". The Washington Post. Associated Press.

[73] Callahan, M.P.; Smith, K.E.; et al. (11 August 2011). "Carbonaceous meteorites contain a wide range of extraterrestrial nucleobases". Proc. Natl. Acad. Sci. U.S.A. 108 (34): 13995–13998. Bibcode:2011PNAS..10813995C. doi:10.1073/pnas.1106493108. PMC 3161613. PMID 21836052.

[74] Steigerwald, John (8 August 2011). "NASA Researchers: DNA Building Blocks Can Be Made in Space"

therefore all matter, including the elements that make up our human bodies are stellar leftovers. I think he meant more than that though, something more poetic or perhaps even magical. Science is finally catching up to Sagan. Neil deGrasse Tyson has paraphrased the sentiment; "We are stardust brought to life, then empowered by the universe to figure itself out." We are now finding out that nebulae are where this process begins.

In the late nineties to early two thousands, much academic work was being done in the astrobiology community to analyze deep space objects for spectral signatures of various complex molecules. A hypothesis had been proposed that as certain types of nebulae approach the ends of their lives, convection currents would cause carbon and hydrogen from the nebulae's core to get caught in stellar winds, and become radiated outward.[75] [76] Upon leaving the hot core, the cooling process would theoretically allow atoms to bond to each other in a variety of ways and eventually form complex particles of upwards of a million atoms, including a highly structured and geometrical class of complex organic molecules known as polycyclic aromatic hydrocarbons (PAHs).

PAHs are a kind of precursor to the more complex proteins that are responsible for metabolism and structure

[75] L. Allamandola, D. Hudgins, S. Sandford (1999) "Modeling the unidentified infrared emission with combinations of polycyclic aromatic hydrocarbons". Physics, The Astrophysical journal, 1999
[76] Mulas, G.; Malloci, G.; Joblin, C.; Toublanc, D. (2006). "Estimated IR and phosphorescence emission fluxes for specific polycyclic aromatic hydrocarbons in the Red Rectangle". Astronomy and Astrophysics. 446 (2): 537–549. arXiv:astro-ph/0509586. Bibcode:2006A&A...446..537M. doi:10.1051/0004-6361:20053738. S2CID 14545794.

in living things, including DNA. Thus it is presumed that early life arose directly from such molecules. One interesting thing about PAHs is that they only occur naturally in fossil fuels, or as a result of burning organic material. They do not form in nature except as a result of biomass decay. We can not form them in the lab without modifying existing biomass. We have discovered one place however, where PAHs are spontaneously generated without the requirement for pre-existing biomass to process. That place is inside a nebula.

PAHs were first observed in 2004 inside the Red Rectangle Nebula,[77] with similar results following from other nebulae[78] and protostar locations.[79] The chemicals identified included sugars specific to the manufacturing of RNA, and fullerenes (aka "buckyballs"). RNA works in conjunction with DNA to encode protein manufacture in biological entities. This finding suggests that complex organic molecules may form in stellar systems prior to the formation of planets, eventually arriving on young planets

---

[77] Mulas, G.; Malloci, G.; Joblin, C.; Toublanc, D. (2006). "Estimated IR and phosphorescence emission fluxes for specific polycyclic aromatic hydrocarbons in the Red Rectangle". Astronomy and Astrophysics. 446 (2): 537–549. arXiv:astro-ph/0509586. Bibcode:2006A&A...446..537M. doi:10.1051/0004-6361:20053738. S2CID 14545794.

[78] García-Hernández, D. A.; Manchado, A.; García-Lario, P.; Stanghellini, L.; Villaver, E.; Shaw, R. A.; Szczerba, R.; Perea-Calderón, J. V. (28 October 2010). "Formation Of Fullerenes In H-Containing Planetary Nebulae". The Astrophysical Journal Letters. 724 (1): L39–L43. arXiv:1009.4357. Bibcode:2010ApJ...724L..39G. doi:10.1088/2041-8205/724/1/L39. S2CID 119121764.

[79] Than, Ker (August 29, 2012). "Sugar Found In Space". National Geographic

early in their formation.[80] Or as astronomer Letizia Stanghellini summarized, "It's possible that buckyballs from outer space provided seeds for life on Earth."[81]

Transformers

Even in the vast spaces between galaxies and nebulae, scientists have discovered a background of cosmic dust that contains organic molecules that are larger and more complex than PAHs.[82] In 2012 and 2013 two groups of scientists at NASA and at the Atacama Large Millimeter Array[83] (ALMA) in Chile reported that PAHs formed within nebulae may be transformed, through hydrogenation, oxygenation and hydroxylation, into more complex organics after emerging from the nebulae and

---

[80] Jørgensen, J. K.; Favre, C.; et al. (2012). "Detection of the simplest sugar, glycolaldehyde, in a solar-type protostar with ALMA" (PDF). The Astrophysical Journal. eprint. 757 (1): L4. arXiv:1208.5498. Bibcode:2012ApJ...757L...4J. doi:10.1088/2041-8205/757/1/L4. S2CID 14205612.

[81] Atkinson, Nancy (27 October 2010). "Buckyballs Could Be Plentiful in the Universe". Universe Today.

[82] Kwok, Sun; Zhang, Yong (26 October 2011). "Mixed aromatic–aliphatic organic nanoparticles as carriers of unidentified infrared emission features". Nature. 479 (7371): 80–83. Bibcode:2011Natur.479...80K. doi:10.1038/nature10542. PMID 22031328. S2CID 4419859.

[83] Loomis, Ryan A.; Zaleski, Daniel P.; Steber, Amanda L.; Neill, Justin L.; Muckle, Matthew T.; Harris, Brent J.; Hollis, Jan M.; Jewell, Philip R.; Lattanzi, Valerio; Lovas, Frank J.; Martinez, Oscar; McCarthy, Michael C.; Remijan, Anthony J.; Pate, Brooks H.; Corby, Joanna F. (2013). "The Detection of Interstellar Ethanimine (Ch3Chnh) from Observations Taken During the Gbt Primos Survey". The Astrophysical Journal. 765 (1): L9. arXiv:1302.1121. Bibcode:2013ApJ...765L...9L. doi:10.1088/2041-8205/765/1/L9. S2CID 118522676.

entering interstellar deep space. This was found to occur when PAHs came into contact with interstellar "ice grains" under the effects of cosmic radiation.[84] While these newly minted organics are a step closer to DNA, they are still not as complex as the proteins and amino acids that are considered strictly biological. The quest for life in space continues. Yet, Anthony Remijan, one of the ALMA scientists, concurred with Stanghellini statement, saying that "finding these molecules in an interstellar gas cloud means that important building blocks for DNA and amino acids can 'seed' newly formed planets with the chemical precursors for life."[85] Furthermore, such seeds are now thought to be highly abundant and widespread. By one estimate, these materials may account for more than 20% of the carbon in the universe.[86] It may be that we are coming close to proving Crick correct in his belief that life did not begin on earth but was brought here.

Space Viruses

We have found the basic building blocks for life scattered widely across the universe. We have found what

---

[84] Gudipati, Murthy S.; Yang, Rui (September 1, 2012). "In-Situ Probing Of Radiation-Induced Processing Of Organics In Astrophysical Ice Analogs – Novel Laser Desorption Laser Ionization Time-Of-Flight Mass Spectroscopic Studies". The Astrophysical Journal Letters. 756 (1): L24. Bibcode:2012ApJ...756L..24G. doi:10.1088/2041-8205/756/1/L24. S2CID 5541727.

[85] Finley, Dave (2013-02-28). "Discoveries Suggest Icy Cosmic Start for Amino Acids and DNA Ingredients". The National Radio Astronomy Observatory. Nrao.edu. Retrieved 2018-07-17.

[86] Hoover, Rachel (February 21, 2014). "Need to Track Organic Nano-Particles Across the Universe? NASA's Got an App for That". NASA. Retrieved February 22, 2014.

appear to be the remains of bacteria in asteroids and comets. So far, we have not detected any remains of more complex life forms in space. That is not to say however that there is no evidence. From one perspective, the next step up from bacteria are viruses. Could there be viruses in space? According to NASA's interdisciplinary Virus Focus Group, "space viruses may be "terrifyingly common".[87] NASA is taking the possibility very seriously and have a special task force dedicated to investigating such focus topics as the as-yet-unknown origin of viruses, and the question of "how viruses may have influenced the origin and evolution of life here on Earth, and perhaps elsewhere in the Solar System".

Although viruses outnumber all other forms of life on Earth[88], we know surprisingly little about them. Viruses evolve very rapidly and are subject to extreme rates of mutation. In terms of the "fossil record"[89] though, they seem to pre-date the divergence of life and had infected[90] Earth's last universal common ancestor.[91] In other words they have been on Earth as long as life has been here. Even

---

[87] https://astrobiology.nasa.gov/nai/media/medialibrary/2019/10/StedmanAstrovirologyIntro.pdf

[88] https://massivesci.com/articles/extraterrestrial-life-virus-nasa/

[89] To be clear, viruses do not form fossils in the traditional sense, because they are much smaller than the finest colloidal fragments forming sedimentary rocks that fossilize plants and animals. However, the genomes of many organisms contain endogenous viral elements. These DNA sequences are the remnants of ancient virus genes and genomes that ancestrally invaded the host germline. These sequences provide retrospective evidence about the evolutionary history of viruses, and have given birth to the science of paleovirology.

[90] Mahy, W.J.; Van Regenmortel, MHV, eds. (2009). Desk Encyclopedia of General Virology. Academic Press. ISBN 978-0-12-375146-1.

[91] https://en.wikipedia.org/wiki/Last_universal_common_ancestor

so, it is still unclear if viruses can be truly considered to be "living". They are in many ways more similar to an artificial life form. Viruses are not made out of cells, they don't grow or metabolize. They only can do one thing, release their DNA into the host.

There are several competing theories regarding the origin of viruses, one of which states that viruses evolved from complex molecules of protein and nucleic acid before cellular life first emerged.[92] According to this theory, viruses then contributed to the rise of cellular life[93] by contributing to the gene pool.

Although we have not yet found direct evidence of viruses in space, there is a fair bit of anecdotal evidence linking viral outbreaks with incoming cosmic materials in the form of meteorites.

In 2007, a bright, flaming object streaked through the sky, trailing a plume of dark smoke, and struck the earth near the village of Carancas, Peru[94] releasing what witnesses described as a noxious vapor. A small crowd of locals gathered to look at the crater, many of whom soon became sick, presumably due to the fumes. It was initially proposed that poisoning from heavy metals including arsenic compounds from the groundwater were the likely

---

[92] Villarreal, L.P. (2005). Viruses and the Evolution of Life. ASM Press. ISBN 978-1555813093.

[93] Nasir, Arshan; Kim, Kyung Mo; Caetano-Anollés, Gustavo (2012-09-01). "Viral evolution". Mobile Genetic Elements. 2 (5): 247–252. doi:10.4161/mge.22797. ISSN 2159-2543. PMC 3575434. PMID 23550145.

[94] https://en.wikipedia.org/wiki/2007_Carancas_impact_event

causes,[95] however, experts were unable to agree on this diagnosis, and other explanations quickly aired. One such alternate hypothesis was that the meteorite may have contained viruses or bacteria from space. The meteorite originated in the asteroid belt, where we now know bacteria are certainly possible, so this was not as far-fetched as many vocal skeptics would have us believe. Still, no consensus could be reached, and both medical professionals and scientists involved in the investigation were stymied by the "mystery illness".

There is a central question here that should be addressed, and to which there appears to be a lack of consistent, publicly available data. While all reports agree that the total number of illnesses was approximately two hundred, what is not clear in the news stories is whether these people all had visited the impact site, or whether there may have been a component of contagion in the outbreak. This of course is a vital question because while a biological infection would likely be at least somewhat contagious, environmental poisoning is definitely not. We do know that a town hall meeting the evening of the impact was attended by over eight hundred villagers and that these attendees were warned not to approach the site. This indicates a strong possibility of contagion playing a role, but it is by no means conclusive without further data.

Unfortunately, there is no clear answer on this case, but thankfully, there are many other cases to consider. Throughout history, epidemics have often been portended

---

[95] José Orozco, National Geographic News, "Meteor Crash in Peru Caused Mysterious Illness", September 21, 2007. Retrieved October 10, 2007.

by comet or UFO sightings. Again, we will likely have difficulty making conclusions due to data scarcity, but due diligence requires us to examine all the evidence to the best of our ability.

One very well-known case is that of the so-called "Miracle of the Sun"[96] which occurred in Fatima, Portugal in 1917 and was immediately followed by one of the most deadliest pandemics the world has ever seen — the infamous Spanish Flu epidemic which killed fifty million people.

In 1916 and 1917, a group of children claimed to have been visited by what they called variously an angel or the virgin Mary who told them that October 13th 1917 she would appear publicly. Despite the inevitable skepticism, and a brief arrest of the children, somehow or other, their story became widely believed and spread until the national newspapers covered the story and thousands of devout Catholics pilgrimaged to Fatima to witness the promised event, so that on the appointed day there were at least thirty thousand, or by some estimates up to one hundred thousand[97] devotees present and staring at the sky.

Exactly what these throngs of witnesses did see was described in a variety of ways. Many said that the sun was replaced by a spinning disc in the sky which cast multicolored light beams across the landscape. Some saw the sun careen towards the Earth or zig-zag across the sky. The phenomenon was witnessed not only by the gathered

---

[96] https://en.wikipedia.org/wiki/Miracle_of_the_Sun

[97] Donal Anthony Foley. Marian Apparitions, the Bible, and the Modern World. Gracewing Publishing; 2002. ISBN 978-0-85244-313-2. pp. 404–.

thousands, but also by independently corroborated reports of persons located more remotely, some up to eighteen kilometers away.

Interestingly, at no point during the ten minute encounter did the object appear to shapeshift into a virginal form, nor to speak to the assembled masses. Nor did the object crash into the earth or break up in the atmosphere. If we take the witness descriptions at face value, this was neither an angelic messenger, not a meteorite. So what was it? Some theorists assume it had to have been a UFO. In a strictly literal sense, it was. At least it appeared to be a flying object, and though many tried to identify it, its ultimate identification remained a mystery. Theorists also assert that there was a correlation between the events of Fatima and the start of the so-called Spanish Flu epidemic of the following year. Again, strictly speaking, there may be a correlation, but this does not imply causation, and even the correlation itself is tenuous because we don't have good data as far as when the epidemic actually began. The world was embroiled in the throes of "The Great War", with governments doing their level best to control the flow of information and propaganda. Tremendous political and social forces were at play which likely suppressed news of national outbreaks until such point as it became uncontainable. On the other hand, there is also evidence that the pandemic may have started as early as 1916, well before the Fatima phenomenon. It must be noted though, that the Fatima phenomenon itself began in early 1916 when the three children experienced their first angelic encounter. So perhaps correlation has been established to a fair degree, but what of causation? Nowhere in the Fatima story is there any hint of divine retribution being prophesied, nor even taken credit after the fact.This does not appear to be a case of God smiting the evildoers. But if not God, then who? Could some other extraterrestrial force be behind both the Fatima encounter and the pandemic? All we can say is that the evidence is inconclusive on that question.

Claims have been made that a wave of UFO sightings in 1731 preceded the plague of 1732. In Sheffield, England, a witness described "a dark red cloud, below which was a luminous body which emitted intense beams of light." In Kilkenny, Ireland a witness reported a great ball of fire that shook the entire island and the whole sky seeming to burst into flames. Meanwhile in Romania, there appeared "a great sign in the sky, blood red and very large." Finally, in Italy, "a luminous cloud was seen in the sky" which disappeared over the horizon. These claims all occurred within a twenty four hour period. There is little evidence to corroborate a claim to a coincident cause of plague, for in actual fact, numerous plagues were occurring globally both before and after these events, for many years, and on a global scale.

The unfortunate fact is that plagues and epidemic occurred ubiquitously and regularly throughout history and despite claims that the black plague was started by mists being sprayed from flying bronze ships, or the Justinian plague coinciding with glowing gold shields and bronze ships in the sky, the truth is that any solid evidence is frustratingly hard to come by and these claims remain tenuous at best.

I would be remiss not to mention the possibility of extraterrestrial sources for some Biblical plagues described during the exodus and the conquest of Canaan. These are discussed in my book "UFOs In The Bible" According to UCLA professor Richard Rader, "in the mythological imagination, plagues are always the result of the gods being angry or upset with us." Are these gods merely beings from another region of space? Or are the gods the actual microbes themselves? Perhaps viruses are the intelligent life we have been searching for, or perhaps they are a terraforming tool, or an ongoing program of DNA insertion missions. If so, were they intended as weapons, or colonization missions, or are they purely accidental "hitchhikers"? Has all of Earth's biological material, DNA, and even life itself been injected to earth from comets, as proposed by brilliant scholars such as Fred Hoyle, Chandra Wickramasinghe, and others?[98] In my opinion this is certainly plausible, and based on a lot of albeit somewhat circumstantial evidence, maybe even probable, but we are going to require a lot more evidence of a more conclusive nature before we can come anywhere close to proving it.

---

[98] Edward J. Steele, Shirwan Al-Mufti, Kenneth A. Augustyn, Rohana Chandrajith, John P. Coghlan, S.G. Coulson, Sudipto Ghosh, Mark Gillman, Reginald M. Gorczynski, Brig Klyce, Godfrey Louis, Kithsiri Mahanama, Keith R. Oliver, Julio Padron, Jiangwen Qu, John A. Schuster, W.E. Smith, Duane P. Snyder, Julian A. Steele, Brent J. Stewart, Robert Temple, Gensuke Tokoro, Christopher A. Tout, Alexander Unzicker, Milton Wainwright, Jamie Wallis, Daryl H. Wallis, Max K. Wallis, John Wetherall, D.T. Wickramasinghe, J.T. Wickramasinghe, N. Chandra Wickramasinghe, Yongsheng Liu, Cause of Cambrian Explosion - Terrestrial or Cosmic?, Progress in Biophysics and Molecular Biology, Volume 136, 2018, Pages 3-23, ISSN 0079-6107, https://doi.org/10.1016/j.pbiomolbio.2018.03.004.

Space Tardigrades

There may or may not be viruses living in space. However, there is a very strong probability of even more complex life forms currently surviving in space. You see, a couple years ago we put tardigrades on the moon.

In 2019, Israel Aerospace Industries' "Beresheet" lunar lander crashed on the moon.[99] Shortly thereafter, scientists involved in the mission revealed that a capsule containing several thousand tardigrades had been aboard and that these organisms may possibly have survived the crash. Tardigrades are a microscopic animal that is typically found living in ponds or in the dirt on Earth. They are peaceful, non toxic, and in no way harmful to humans. They also possess the unique and most uncanny ability to survive being completely dehydrated, frozen, and subjected to conditions normally considered deadly, and to return to life as normal when rehydrated. On previous space missions, tardigrades have been exposed to the hard vacuum of space and survived.[100] The Israeli authorities were quick to assure reporters that "there is no real danger they will spread across the Moon"[101] but given their propensity to live through nearly any circumstances, and coupled with the fact as examined earlier that the moon

---

[99] Lidman, Melanie. "Israel's Beresheet spacecraft crashes into the moon during landing attempt". The Times of Israel.

[100] Resnick, Brian (6 August 2019). "Tardigrades, the toughest animals on Earth, have crash-landed on the Moon - The tardigrade conquest of the Solar System has begun". Vox. Retrieved 6 August 2019.

[101] Oberhaus, Daniel (5 August 2019). "A Crashed Israeli Lunar Lander Spilled Tardigrades on the Moon". Wired. Retrieved 6 August 2019.

does have water and a very small amount of oxygen available, this author remains hopeful that this accidental first terrestrial moon colony will indeed flourish. They might end up to be the first wave in a series of terraforming operations on good old Luna.

Life On Mars

In 1976 NASA's Viking mission landed on Mars with the intent of mapping, imaging, atmospheric sampling, and geological research. Even though no one expected to find any positive results, a few ancillary experiments were included whose purpose was to detect potential biosignatures - just in case. There was really no expectation that anything might be alive on that barren Martian surface. It was almost more about ruling it out. Three separate experiments were designed to detect micro-organisms in the Martian soil. In the first experiment, radioactive nutrients were injected into the soil with the idea that if the ground were "biologically active", the nutrients would prompt latent metabolic processes ending up with the release of traceable radioactive carbon dioxide. A second similar experiment consisted of injecting irradiated carbon dioxide in the hopes that any photosynthesizing life (aka, plants) would result in traceable oxygen being released. Thirdy, water was injected into the soil, which should yield oxygen production, again indicating photosynthesis.

The results of all three of these experiments indicated the presence of living organic material. On the first ever Mars landing, way back in 1976. Shockingly, we found life on Mars nearly fifty years ago, and no one is talking about it. The engineers and biologists who designed these experiments (keeping in mind these are some of NASA top people) considered the results of the experiments incontrovertible evidence of life. Apparently someone somewhere higher up decided this did not fit the agenda, and the results were swept under the rug. Not to be dissuaded, the two primary leads responsible for the experiments, Gil Levin and Patricia Straat, remained confident that bacteria had in fact been detected on Mars, a result they stood by until their respective deaths in 2021 and 2020. There is life on Mars.

Yet, NASA successfully avoided announcing the findings and seems to have purposefully avoided any further investigations along these lines for the next four decades. That may soon be changing. In January 2014, NASA reported that current studies on the planet Mars by the Curiosity and Opportunity rovers will now be searching for evidence of "ancient life", and environments that may have supported life at one time. The search for evidence of habitability, fossils, and organic carbon deposits on the planet Mars is now a primary NASA objective.[102] They are still hedging their bets and have not admitted their earlier coverup, but at least it is a step in the right direction. They are probably afraid of being outed when someone like Elon Musk beats them to the punch.

---

[102] Grotzinger, John P. (24 January 2014). "Introduction to Special Issue – Habitability, Taphonomy, and the Search for Organic Carbon on Mars". Science. 343 (6169): 386–387.
Bibcode:2014Sci...343..386G. doi:10.1126/science.1249944. PMID 24458635.

Early Earth

Prior to around two billion years ago, it is thought that the early Earth may have had what chemists would call a "reducing atmosphere". Without going too deeply into the chemistry, I think it's at least somewhat accurate to simplify this to mean essentially, that early earth was not rich in oxygen as it is now. Although this may sound like a very inhospitable environment, we must be cognizant that our perceptions regarding oxygen are highly and strongly biased. Like all animals, we depend on oxygen to live and breathe.

Oxygen, in and of itself though, is highly corrosive. So much so that the term "oxidize" derives from the element's name. To oxidize is to break down another element or compound by simply exposing it to oxygen. Synonyms include 'rust', 'tarnish', 'decompose' and even 'burn'. Yes, fire is nothing more than a rapid oxidation. We tend to think of fire as a destructive force — one that can be harnessed and may prove useful, but destructive nevertheless. This destructive combustion is what drives our internal metabolism. We are literally walking furnaces. But life on Earth wasn't always like this.

The early atmosphere of Earth was essentially devoid of oxygen. and that was good news for the first inhabitants of our planet. It is thought that cyanobacteria evolved in this reducing atmosphere using photosynthetic metabolic pathways. In this sense, these first earthlings were similar to plants, in that they took in carbon dioxide, and used photosynthesis to fuel their cellular activity, expelling oxygen as a byproduct. This environment was not very different from that of interplanetary space.

Space is also considered a reducing environment, due to the presence of solar wind, which consists primarily of hydrogen plasma. Any planetoid with a sufficiently sparse atmosphere also falls into the same category. This includes present conditions on our moon and on Mars. Although today's low oxygen levels on these planets are not conducive to animal life, they are actually quite amenable for photosynthesizing bacteria.

Unfortunately for our photosynthetic friends, their own waste product gradually began to build up, increasing the oxygen levels in Earth's atmosphere, and over the course of maybe a million years or more, they found themselves choked out by their own products. Before they knew it, the atmosphere had changed from a reducing one to an oxidizing one[103], and metabolic processes reversed, killing every living thing on the face of the earth. At least that is the theory, and the simplified worst case at that. In reality, though this narrative is generally true, there were likely exceptions, thanks to the always-inventiveness of life. The sudden injection of toxic oxygen into an anaerobic biosphere did indeed cause the extinction of many existing anaerobic species on Earth,[104] and this event can be considered a mass extinction.[105] However, it is thought that some species managed to survive by modifying their

---

[103] https://en.wikipedia.org/wiki/Great_Oxidation_Event

[104] Hodgskiss, Malcolm S. W.; Crockford, Peter W.; Peng, Yongbo; Wing, Boswell A.; Horner, Tristan J. (27 August 2019). "A productivity collapse to end Earth's Great Oxidation". PNAS. 116 (35): 17207–17212. Bibcode:2019PNAS..11617207H. doi:10.1073/pnas.1900325116. PMC 6717284. PMID 31405980.

[105] Plait, Phil (28 July 2014). "Poisoned Planet". Slate.

metabolisms to tolerate the presence of oxygen. These may be the predecessors of all multicellular life forms.

It is fascinating to consider that humans are not the first lifeforms on earth to have dramatically and irrevocably modified the environment, leading to their own extinction. We may not be the last to do so either. It was only after this atmospheric "flip" that the evolution of aerobic respiration became possible. The death of earth became the opportunity for a new type of life. We will never fully comprehend exactly what was lost. We are only now beginning to comprehend the types of lifestyles, and dare I say communities, that are present underneath our noses. Fungi are possessed with intricate communication protocols,[106] and even viruses appear to be communicating amongst themselves with a limited set of chemical signals.[107] Were these anaerobic entities capable of communication? Did they constitute Earth's first civilization? Was Earth once host to a global anaerobic bacterial civilization whose own success was their ultimate demise? Many people see this as our own inevitable fate, but if it can happen once, it can happen again, and it could have already happened. And if it happened on earth, might it have happened elsewhere?

Are They Here?

We come now to the point of our discussion which to me seems most paradoxical. There has long been a rich

---

[106] https://www.smithsonianmag.com/smart-news/mushrooms-may-communicate-with-each-other-using-electrical-impulses-180979889/
[107] https://magazine.scienceconnected.org/2017/02/viruses-talking-peer-pressure/

and vibrant community of scientists and enthusiasts searching tirelessly for any sign of life "out there" but in large part, this community seems to tend toward ignoring the wealth of evidence from another group — those who are quite convinced that "out there" may be the wrong direction to look, because the extraterrestrial life has already come to us — the community of ufologists, UFO witnesses, experiencers and abductees.

Repeatable Observation, Observational Variation, And Observational Bias

Science, and specifically the scientific method, is built upon certain principles, one of the most fundamental is that of repeatability of observation. That is, for any observation to be valid, it must not vary, but be repeatable over a given set of criteria, which may include environmental factors, but which ought not to depend upon the perhaps whimsical expectations of the observer. When the apple dropped upon Sir Isaac Newton's head, he was able to describe the effects and could have dropped a similar apple upon the head of any random passerby with a similar set of effects. When a meteorite shoots through the clouds, there should be general agreement between witnesses on what it looked and sounded like. When chromosomes are observed undergoing mitosis, they do so in a predictable way, even if the fine grained results may not be completely predictable. This is true regardless of what the observer happens to think about the process, or whether or not there is an observer present at all.

Similarly, modern UFO phenomena tend to occur in a variety of flavors, but this variety is by no means infinite.

In fact, there are really only a handful of differences that crop up when examining the testimonies of UFO witnesses and ET experiencers. These phenomena are remarkably repeatable. Of course, we can not control whether or when such phenomena might take place, but the same is true of comets, earthquakes, and pregnancies. Would any man be so stupid to assume pregnancy does not exist, just because it has never happened to him? Yet this is often the type of logic that is turned against the UFO phenomenon. Hundreds of years ago, male scientists and doctors began approaching human health with a scientific mindset, and eventually discovered that pregnancy tends to proceed in a reasonably predictable pattern, resulting in birth. These doctors were somehow even able to determine the cause of pregnancy. Yet, science in the twenty-first century is still embracing an oblivious denial coupled with stubborn refusal to rationally approach the UFO phenomenon.

My hope and assumption is that most readers of this book are not so fearful as to continue this staunch conservatism, but are open minded enough to address a controversial subject in an unabashed and curious way. Let us therefore examine the evidence at hand and ask if it meets the criteria of repeatable observations over a large sample size.

Every year, there are thousands of reported UFO sightings around the world. In Canada alone, the annual average is around one thousand. In the USA, that number is from four to six times higher, depending on whose statistics you go with. Statistics from other countries have been difficult to track down, but there are certainly many documented UFO cases from all over Europe, Africa, Asia,

South America, and Australia. Are there any consistently repeated factors within these reports?

UFOs do not come in all shapes and sizes. There are a very limited set of configurations that are widely reported around the world:

- Saucers
- Triangles and V-shapes
- Ovals and "Tic-Tacs"
- Long rods or pillars or "cigars"
- Large looming shapes with lots of lights
- Orbs

Likewise when it comes to alleged alien species that witnesses have described, again there are a limited number of options:

- The Greys
- Nordic/Plejaren/Pleiadeans
- Reptilians
- Basically Human

Within any dataset of this size there are bound to be outliers and anomalies, and of course this observation applies to the UFO phenomenon, but the bulk of the cases do appear to fit the mold of these small sets of observational types. We shall see this directly later on when we look at some example sightings.

A theory regarding observations has been put forward, that can be boiled down to the idea that UFOs are merely a mental construct that has been manifested somehow (so far quite inexplicably) by the observer. Proponents of this theory claim that there are paranormal phenomena that seem to be related or coincident. Further, they state that variances in how these phenomena are perceived is informed by the cultural milieu in which they are experienced. That is, according to this theory, observations of *mechanical* UFOs, ships, saucers, etc., became common only after the invention and general introduction to the populace of similar mechanical objects such as planes and cars. Prior to the mechanical age, it is claimed, paranormal activity was generally described in terms of monsters such as dragons or fairies.

There is certainly a grain of truth in this theory. Before humans had any experience with mechanical machines, there was a lack in the available language to describe such things. This fact alone does not necessarily transfer to the observation itself, but rather, only to the means of describing the observation after the fact. This is something that I examine in detail in my book *"UFOs In The Bible"*, specifically as it relates to the encounters of the prophet Ezekiel and the patriarch Moses.

In reality though, the descriptions of mechanical objects can be easily interpreted as such even by "primitive" observers, although some of the motifs may vary slightly. For example, the word "chariot" was often used in the ancient world, where a modern observer may say "ship". Both are metallic mechanical transportation devices. Similarly, the types of metals known during different eras have influenced the description in obvious ways. A modern observer may see a 'steel' or 'aluminum', or even a 'plastic' craft, whereas an ancient observer tends to describe it as 'bronze', 'copper', or 'gold'. The point is, it's metallic and shiny. No one is claiming to have run metallurgical studies and elemental analysis on the thing. Metallic flying mechanical objects have been described for thousands of years, long before the invention of the modern metal airplane.

Modern accounts of glowing orbs, and of diminutive extraterrestrials draw similarities between many ancient traditions such as the fae, or various other "little people" in legend. We ought not to assume that we understand these traditions, based on some bastardized Disney version produced in the sixties. There is a richness and a sense of rationalism underlying many of these traditions in the context of their native cultures that we can not truly appreciate from our biased viewpoints and our own indoctrination that they are "just stories". Many of them seem to parallel quite closely many elements of ufology.

I will admit it is more difficult to answer for the seeming discrepancy on the part of dragons. Where are the dragon parallels in modern ufology? They seem mainly obscured. Mainly, yes. Entirely, no. Nasa astronaut Franklin Story Musgrave, saw something snakelike, as will be discussed on page 87.

The framework of repeated observation brings with it an inherent problem; that of observational bias. That is, many observers will jump to conclusions that match their expectations rather than their actual observed details. The human mind excels at filling in the blanks. Many UFO witnesses just happen to be UFO enthusiasts. This can be partly explained by the fact that an enthusiast is more likely to be looking in the right place at the right time, however, it appears that many UFO reports are highly speculative, based on very scant data, but influenced by the desire of the observer to see a UFO. This is especially true in many of the thousands of cases consisting of reports of tiny lights in the night sky which could very easily be mundane aircraft, satellites, comets, or other manmade or natural phenomena. This is far less likely to be the case however, when it comes to close encounters. In a close encounter, there is discernible shape and size, which rule out most simple explanations. In these cases, it doesn't seem to matter much whether the witness is a willing believer or not; in fact many self described skeptics have had their minds changed, and in some cases, their lives dramatically altered, by their UFO experiences. Therefore it is difficult to draw a clear correlation between the UFO phenomenon as a whole and the expectations and biases of the observers. Sadly, the same cannot be said about the UFO skeptics.

Trashing The Evidence

Skeptics of all sorts tend to have well formed biases. They often have something to prove and enjoy a good argument. I should know. I am one.

This is not necessarily a problem... until it is. A good skeptic will hold their ground until proven wrong. The problem is that it is difficult to prove a thing when all evidence is thrown out the window. This tends to be true in the UFO debate. A good ufologist will acknowledge the biases present in the community and the resulting reports and so-called 'evidence'. A generous quantity of "grains of salt" must be used when investigating UFOs. Despite this, it is not at all difficult to weed out spurious data and find reports that appear to be highly substantial and reliable. Unfortunately, on the side of the skeptics, all too often there is little willingness to sort the data. Instead, the baby is thrown out with the bathwater. If you think about it, this makes a lot of sense. I am not a sports fan. If you ask me about my favorite football or baseball team you won't get much of a conversation from me. It's not because I think all football teams are equally great. It's because I don't care enough to formulate an opinion. I am not interested in wasting my time learning about team stats and players, and I remain purposefully uneducated about the whole matter. The same thing can happen with any controversial topic, and UFOs are no exception. If your friend thinks UFOs are stupid, he is not likely to read books or investigate cases and make up his mind based on data. Instead, he will take what little data happens to float across his radar, and he will quickly discover the truth that a lot of that data is unsubstantiated, hyper-sensationalized nonsense. Based on that data alone, his conclusion is correct: UFOs are stupid. They may be stupid but it doesn't make them any less real. Except to your friend. Since UFOs are stupid, they can't be real and are not worth wasting time talking about, letting alone reading about. By definition, this is a foregone conclusion.

Many scientists fall into this trap. For a variety of reasons, some of which are quite valid, highly intelligent and highly trained people specializing in a certain field may tend to view other fields as "beneath them". This may be even more true of "nonscientific pursuits", or worse yet "pseudoscience". Sadly, they then ignore the real evidence — the hard data, often writing it all off as "anecdotal evidence", a term used to belittle anything they can not objectively measure. However, the truth is that there are whole fields of science built around anecdotal evidence. While a physicist, chemist, or engineer concerns himself entirely with quantitative data, it is *qualitative* methods that fuel the fires of history, psychology, sociology, anthropology, political science, and numerous other disciplines. Quantitative methods are not "less sciency" (although they may be "less mathy"). They are not pseudoscience. Neither is any field that uses well defined methods to examine data. In ufology, the well defined methods are perhaps not spelled out, but they include repeatable procedures such as sorting through extraneous data, looking for reliable data, and identifying trends and patterns in that data. This is exactly what J. Allen Hynek did when he invented his "Close Encounter" scale way back in 1972. This is why we deemed it important to talk about logical fallacies in a previous section. This is the whole point of this book. Ufology is a field that *can* be studied scientifically. Therefore, in my opinion it *must* be studied scientifically. There is no longer room for the inclusion of fallacies, suppression, bullying, or pure unadulterated ignorance in this field.

I understand that we all have our various areas of interest. We have to limit our scope. I for one do not have time or resources to study every field I might find intriguing. I do not expect every chemist or mathematician to dig into all the UFO data themselves. What I demand though, if I may be so bold, is for self proclaimed scientists, both professional and amateur, who have chosen to invest time and energy and education into the question of whether there might be life in the universe, make a conscious and intentional decision to stop ignoring the reams of readily available evidence of that exact same life they claim to be chasing, right here in front of us on good old planet earth.

All Around

According to a poll conducted in 2021, 65% of Americans believe that intelligent life exists on other planets. Of course, belief doesn't make it true, only evidence has that privilege. So what of the four thousand to six thousand UFO sightings reported in the USA every single year? Can these be taken as evidence? Certainly, many of these are simple cases of mistaken identity. How many is not clear. A very conservative estimate may be that up to ten percent are not easily explained by some mundane phenomena. That's still four hundred to six hundred unexplainable cases per year, in the USA alone, to say nothing of the same types of reports happening on a global scale. The officially reported sightings are certainly much fewer than the actual sightings themselves. As observed by UFO lecturers at conferences by means of casual audience polls, the number of attendees who had either witnessed a UFO or experienced extraterrestrial contact, was roughly ten times higher than the number of those who had actually reported their experiences.

Although there have been some recent advances in cultural acceptance, for the most part, UFO witnesses, and particularly ET experiencers, still are subjected to significant derision and mockery. According to Harvard psychiatrist John Mack, the fear of social rejection, humiliation and invalidation can often be more traumatic to an abductee than the actual experience of abduction.[108] The fact remains that as a culture we have a difficult time

---

[108] https://www.gaia.com/article/do-thousands-of-alien-abduction -accounts-add-up

believing anecdotal data from the average American. But what about the stories and opinions of someone who has been to space himself? — Can we take the word of NASA's astronauts?

Several expert astronauts claim to have made observations from orbit that caused them to believe in UFOs and in extraterrestrial life. In 2021 a story about Buzz Aldrin's supposed encounter with aliens was widely circulated on social media. However, Aldrin has denied the story and it appears to have been fabricated by a sensationalist British tabloid. There are plenty of other astronauts though, that do have their own stories to tell.

One of NASA's astronauts during the pioneering Mercury Program, Scott Carpenter, upon looking back over the space program said "At no time, when the astronauts were in space were they alone: there was constant surveillance by UFOs." Carpenter did not provide further details, but several other astronauts have.

American astronaut Leroy Chiao, was commander of the International Space Station in 2005, when he and the entire crew saw an inexplicable V shaped arrangement of lights in space.

In 1991, cosmonaut Musa Manarov, caught a cylindrical UFO on film.[109] The object resembled a short rod, or elongated tic-tac shape.

Astronaut Gordon Cooper, who flew in NASA's Mercury 9 and Gemini 5 missions, first witnessed a UFO while flying a fighter jet over West Germany in 1951. A few years later Cooper was assigned to the top secret Air Force Test Center at Edward Air Force Base. It was here

---

[109] https://www.dailymotion.com/video/x3ztiz8

on May 3, 1957 during testing in the desert that Cooper and two other test center staff, James Bittick and Jack Gettys, witnessed a "strange-looking, saucer-like" aircraft that landed near them, then took off again, all without making a sound. They captured the action on both regular still cameras and on high-speed (for the time) thirty frame per second motion picture film which they sent to the Pentagon upon request. Cooper fully expected that there would be a follow-up investigation, since an aircraft of unknown origin had landed at a classified military installation, but instead, he never heard about the incident again and was never able to track down what became of the photos and the film. It soon became clear to Cooper that the military and the government were covering up information about UFOs. In his memoirs he pointed out that there were hundreds of reports made by his fellow military jet pilots.[110]

In 1969 astronaut Neil Armstrong became the first human to set foot on the moon. According to unsubstantiated rumors, during this mission there may or may not have been a conversation recorded where Neil Armstrong purportedly reported to mission control that there were other spacecraft present on the other side of the crater. I have not been able to ascertain whether this story is legitimate or if it is another hoax.

US Air Force Sergeant Karl Wolfe was an electronics technician with a top secret crypto clearance, who worked with the tactical air command and with the NSA. In 1965 he analyzed photographs taken by the Lunar

---

[110] Gordon Cooper and Bruce Henderson; *Leap of Faith: An Astronaut's Journey in the Unknown*, ISBN: 9781504054249

Orbiter, which appeared to show detailed artificial structures on the surface of the moon.

During the Apollo 9 mission in 1965, astronauts James McDivitt and Ed White reported seeing "something out there" about the shape of a beer can flying outside their ship. It was reported as a UFO, and McDivitt later joked that he had become "a world-renowned UFO expert."

Speaking of world-renowned experts, you would be hard pressed to find a more credible witness than astronaut Franklin Story Musgrave, M.D. Musgrave was a marine, a doctor, an engineer, a mathematician, and an astronaut with six academic degrees. He piloted and crewed six space flights aboard NASA's space shuttles. During two of these missions, he says he saw a "snake" in space. During an interview in 1995 Musgrave explained: "On two of my missions, and I still don't have an answer, um, I have seen a snake out there, six, seven, eight feet long. It is rubbery because it has internal waves in it and it follows you for a rather long period of time." On another occasion, Musgrave told *Omni Magazine*: "On two flights I've seen and photographed what I call the snake, like a seven-foot eel swimming out there."

He summed up by saying, "The more you fly in space the more you see an incredible amount of things out there, and that sort of brings to you, really a certainty, that other living creatures are out there. Some incredibly primitive, more primitive, some just proteins coming together, amino acids and some just single-cell organisms and others civilizations that have been around for a million years that are doing unimaginable kinds of things." Some of this must surely be conjecture. How could Musgrave have seen microscopic organisms in space through the window of a space shuttle, or how could he testify to the age of anything he saw? Nevertheless, he did see something strange that he had no explanation for, flying in space, thus fitting the criteria for a UFO.

As far as Musgrave's comments regarding tiny organisms and ancient civilizations; could this information have been part of the classified astronaut training material? We have already established that NASA has known about life in space and has been keeping it a secret since the first Mars missions, if not earlier. Might NASA know more about this supposed ancient civilization? When you think about it, a civilization is basically a requirement for space flight, and therefore for the existence of any UFOs from elsewhere. In order to develop even the crudest forms of space travel, a very advanced infrastructure must be in place, underpinned by a complex stack of technologies backed by advanced scientific knowledge. The civilization part of the equation is basically a given. But what about the age of "a million years"? Is there any reason to treat this statement as anything more than wild speculation? Actually, yes.

The vast distances involved in space travel would necessitate exorbitant lengths of time spent in travel. This fact is often brought up by skeptics as to why UFOs are impossible. Frankly, it just takes so long that nobody would ever bother to try such a journey. As previously discussed in the section on exoplanets, a trip to the closest habitable planet would take six hundred thousand years. And that's just for the journey itself. Consider also the planning, the building up of technologies to the necessary levels. How long does that take? The generally accepted "mainstream view" is that Homo Sapiens evolved on earth around three hundred thousand years ago. That seems like a reasonable comparison for our theoretical spacefarers. Add that to the travel time and we have nearly a million years, just like Musgrave figured.

There is another line of evidence pointing to his conclusion. That is the fact that UFO phenomena have been reported for a very long time, all throughout recorded history and even earlier, back in the mists of time where the only available information to have survived has inadvertently been transformed into legend and myth. It appears that the UFOs not only take a long time to get here, but they have also already been here for a very long time, and those anecdotal stories have stuck with us, whether we believed them or not.

UFOs Throughout History

There is insufficient space in this book for a full listing of even the most relevant, important, or interesting UFO encounters in the modern record. There have been and will continue to be many other books written to focus on this topic. Instead, we will provide only a sample of each type of craft so that these examples can act as a comparison for earlier reports. The aim of this section is to focus on the abundance of historical claims into antiquity and beyond, by giving the briefest of summaries possible, and to show the similarities — in other words, the *repeatability* of these observations through the ages.

This chapter is a summary of historical UFO observations excerpted from the book *"Before Roswell"* co-authored by Barbara DeLonge and myself, and based primarily on Barbara's lifetime of research. Her full list can be browsed at her website.[111] As Barbara sums up, *"No anomalous phenomena is as enduring through the ages, more witnessed and reported, or more discussed and debated than the UFO phenomenon."*

The celebrated Roswell crash in July 1947 is the most famous case of flying saucers, as well as government mismanagement and cover-up. Reportedly, a flying saucer type UFO crashed on a ranch just outside of Roswell, alien bodies were discovered, and the news sources reported that the government had recovered both the saucer and its crew. The next day, the military issued a report denying their claims from the previous day, and now stating that the debris recovered was from a weather balloon. However, the weather balloon explanation is a clear fabrication, as evidenced by the further top-secret communications and special projects which ensued. A short time later, a letter dated September 23rd, 1947 from Lt. Gen. Nathan F. Twining, Chief of Staff of the U. S. Army to the Commanding General of the Army Air Force established an official study of flying saucers to be called *Project Sign*. The setting up of Project Sign evidently took too long in the eyes of top brass, so it was escalated in priority in a letter dated 30 December 1947 from Maj. Gen. L. C. Craigie. The army demanded action, and it demanded it in a hurry. Obviously they were under no illusions that weather balloons proved such an imminent threat.

With all the publicity and controversy around this case, what many people do not realize is that Roswell was

---

[111] https://barbaradelong.com/special-projects/ufos-in-the-spiritual-realm/the-ufo-conundrum/

not an isolated incident. In 1947 alone, thirty-three UFO sightings were reported. Most of the sightings occurred in the western USA, west of the Rocky Mountains. In fifteen of the cases, objects seen were disc or saucer shaped, six were spheres, and three were cigar shaped. Nine of these cases involved multiple UFOs, ranging from five to twelve in number. Two of the ships crashed. Five cases appear to be indicative of some type of cloaking technology used on the craft. In nine of the cases the witnesses were impressed by the speed and maneuverability of the craft. Three of the incidents included observation of entities outside their ship. In one case the entities injured the witness.

The previous year was one of the most prolific for UFO sightings. In the eight month period from May to December of 1946, close to two thousand UFO sightings were reported in Sweden, Norway, and Finland. Around ten percent were tracked on radar. Most of the objects were cigar-shapes, though a few discs and spheres were reported. Many of them disappeared into lakes. The so-called "Ghost Rocket" invasion of Sweden was well documented, and official investigations took place involving Scandinavian, British, and U.S. military authorities. A number of "ghost rocket" sightings also took place that same year in Greece, Italy, Portugal, and Belgium.

Throughout World War II, hundreds of pilots on both sides of the conflict reported "Foo Fighters," bright, unidentified flying objects that move in the sky in a strange manner. Often described as either swarms of red or green luminous balls or groups of small metallic disks, they followed aircraft or flew around them, giving the impression of intelligent behavior. Some observers saw them touch the wings or the tail assemblies of the aircraft without causing any visible damage, thus raising the question of whether they were material objects at all. Numerous other sightings concerning larger, cigar-shaped, disk-shaped, or sphere-shaped objects were also recorded in both camps. Authorities on both sides of the war took these reports seriously and treated them as a threat, both sides initially assuming them to be secret enemy weaponry or reconnaissance of some kind. President Franklin D. Roosevelt penned a Top Secret memo dated February 22, 1944 on White House stationary for "The special committee on non-terrestrial science and technology." Both the title and the content clearly allude to extraterrestrial life and the document speaks of "coming to grips with the reality that our planet is not the only one harboring intelligent life in the universe."

On the Evening of February 24 1942, The local US Army detachments manned all battle stations as southern California apparently came under attack. A full blackout state of emergency was put into effect, covering a huge area including and around Los Angeles, from Bakersfield south to San Diego and eastward to Boulder City and Las Vegas, Nevada. The military scrambled its fighters and set up anti-aircraft armaments, and the sky was awash with searchlights. Official notice went out on the radio that this was not a drill, and that unidentified aircraft were approaching Los Angeles from the west. Over 1,400 munitions shells were expended. Military officials have been tight-lipped with regards to whether the approaching craft were ever identified as Japanese bombers, or if they remained unidentified.

In very similar fashion to the Roswell case, comes another flying saucer crash in Missouri in 1941, six years before the Roswell incident. The Cape Giradeau police department and some military personnel were at what they thought was a plane crash site, but a witness, Reverend William Hoffman has stated that it obviously was not a plane. There was crumpled aluminum like metal debris scattered over the area and an object that was disc shaped. Huffman also described three small humanoid bodies with large heads, big black eyes, no nose, slit for a mouth, and a form fitting uniforms.

The encounters during the war years and immediately following, can be divided roughly into two groups. Cigar shaped objects in Europe in 1946, and spheroidal or saucer/disc shaped objects prior to 1946 and spanning the globe. There were reports of instrument failure aboard at least one witness vehicle, and one plane was even shot down by a UFO. It should be noted in this case, it was the witnesses who engaged the UFOs into combat maneuvers. A wide range of sizes were reported during this period, ranging from a few inches to two hundred feet in diameter. Some of the craft appear to have been amphibious, either emerging or disappearing into the water.

The thirties included a higher than average percentage of close encounters of the second and third kind, resulting in burnt ground, blistered hands on a witness, and the death of an extraterrestrial by human hands.

In the twenties, both short and tall entities were seen associated with discs and spherical ships. Strange motion was observed, including an apparently undulating craft, and a small object that was observed to follow ground contours and patterns including railroad tracks.

The early twentieth century had its share of UFO encounters, the majority of which were mainly in keeping with the similar shapes of later periods; disk or saucer shaped objects, cigar shaped objects and round or spherical objects. However a few uncommon shapes were reported, including ovals, rectangles, triangles, cruciforms and objects with apparent wings. A number of the craft appeared to be amphibious, either rising from or disappearing into bodies of water. One object was shot down by the German Luftwaft. It was not reported whether materials or occupants were recovered from this crash.

The 19th Century exhibited all the usual shapes and colors of UFO, mainly discs or spheres, but also included several sightings of "pillars or fire" or "whirlwinds" which left a path of destruction in their wake. Oddly, some encounters left behind a gelatinous substance. Winged ships with strange men aboard were witnessed across North America.

Notable 18th century UFO experiences included several objects which were able to split and rejoin, several sightings which appeared coincident with earthquakes. One object reappeared daily for a month. Notable witnesses included Sir Edmond Halley, and the entire British court of King George III.

The Middle Ages saw the usual fireballs, spheres, silver discs, and cigar shaped objects in the sky. Additionally, a number of drum shaped objects, airships, and dragons were seen. Numerous instances recounted airships with personnel aboard. More than once these airships were seen dragging some kind of anchor along the ground, and one of the occupants was asphyxiated while attempting to disentangle it, while another group of them barely escaped a public stoning as per French law. Several accounts spoke of fleets of colorful skyships attacking one another or the unlucky human armies battling below.

During the classical period, a number of Roman and Greek historians, including Dion Cassius. Pliny the Elder, Titus Livy, Julius Obsequens, Cicero, Anaxagoras, and Homer relate the appearance of strange lights in the sky, glowing shields, flying swords, multiple moons and suns, golden flying spheres. Numerous encounters describe young men or maiden dressed in white. Fleets of flying objects again were seen to engage in battle with other flying armies or with human forces. Several prominent figures were witnesses, including Emperor Theodosius, Constantine, and Alexander the Great.

As we move further back in time beyond the classical period, we notice a reduction in records. The remaining encounters seem to be more sparse in time as other records are lost. However, the most memorable and important ones remain imprinted in even the sparsest histories and the pattern shifts markedly to that of close encounters of the third and fourth kind. Even so, there are traces of the more recent patterns: pillars of fire, fiery circles and discs remain as traceable patterns. The evidence remains consistent and repeatable, if not controllable in a lab environment.

Why Are They Here?

If we can believe even just a small percentage of the claims of UFO eyewitnesses, it seems that extraterrestrials have been visiting earth for a long time. But why? What do they have to gain? They do not generally appear to be extracting resources, other than in a few outlying cases where it appears that a craft scooped up a large quantity of water. There is no evidence in the body of eyewitness testimony that they have come for our gold. A significant proportion of contactees describe some sort of physical contact in the form of what may best be described as medical procedures, usually but not always fairly benign such as small tissue samples or the like which implies some type of experimentation. However, the vast majority of UFO encounters involve no direct contact with any alien entities. Usually, we appear to be witnessing a "drive by" for unknown reasons and motivations. All of these factors seem to indicate, when taken together, that whoever they are, they are primarily here to observe.

The concept of extraterrestrial watchers is one of the oldest motifs in the human story. As far back as we can trace, our ancestors have included explanations of beings not from earth, whose role is to observe Earth and her inhabitants. Some of the tales take a protective bent, whereas others focus on the ones who have come teaching ancient wisdom and technologies such as fire or language. In other instances, there appear to be biological interventions in the form of breeding or other manipulations of the gene pool.

Perhaps the most clear example of this motif is presented in the book of Enoch, an ancient Hebrew document. The book describes ten thousand "watchers" who seem to be somewhere nearby, but not on Earth itself. Of this group, two hundred members are said to have landed on Earth in a premeditated, but highly illegal, plan to mate with human women, who then bore half-breed children. These two hundred former watchers were banished to earth where they taught their half-human offspring such technologies as astronomy, metallurgy, medicine, and the use of entheogenic psychedelics. This story, or at least some of its elements, is closely paralleled in many other myths from around the world. Many cultures refer to them as "star men" or "the ancestors".

In the Hebrew version, the visitors are ultimately punished and attempts are made to wipe out the unauthorized bastard children. However, these attempts failed to eradicate the species, as acknowledged by the Hebrew book of Genesis, which states in chapter six that they were there "in those days and afterwards" indicating their presence both before and after the supposed ethnic cleansing. The Sumerian legends tell a similar tale, except that these sky people, or "Anunnaki" as they are called in Sumerian, are seen as the heroes of the story, rather than the outlaws. They are acknowledged as coming from somewhere beyond earth, and bringing civilizing technologies within, as well as the creation of humanity itself. Elements of these archetypes resound in the oral histories from South America, to China, from Africa to Australia. Also common to these tales is some kind of disagreement between the watchers themselves. There are those who get rather "hands on" and there are those who maintain that their proper place is to merely observe. This apparent contradiction within alien society may explain one of the apparent paradoxes that faces those involved in the search for extraterrestrial life — that of the apparent radio silence.

Radio Silence

Much effort has been and continues to be put into the act of listening for radio signals from the stars, or more precisely, not from the stars themselves, but from any intelligent inhabitants out amongst those stars. In principle, this idea is fundamentally flawed. Radio waves have been immensely useful to humans here on Earth in the last hundred years or so. Even here though, we constantly run across the major shortcoming of any radio based telecommunication — the problem of dissipation. Like any wave spreading in a medium, the waves must widen as they go. Simple geometry dictates that concentric radiating circles get larger as they move away from their source. Thus, any electrical signal drops off in power as it gets further from its transmitter. Because the circumference of a circle is related to the radius by the formula $C=2\pi r$, the power available in the signal drops off six times faster than the distance from the transmitting antenna. Even here on earth, the effect is very problematic, which is why you can't connect to your bluetooth devices or your home wifi, unless you are fairly close in proximity. It's also why the television networks were invented. Electromagnetic signals simply can't travel great distances before becoming too weak. The networks had to keep rebroadcasting the signal from every major center.

In Earth's atmosphere, the problem is simplified to an approximation of a two dimensional surface, due to the thin atmosphere as compared to the size of the planet itself. Our atmosphere is essentially a thin membrane wrapped around a ball. It is analogous to the rubber sheet of an inflated balloon, basically two-dimensional. Space however is three dimensional. The formula for a two dimensional circumference no longer applies. Instead, radio waves in space propagate through the three dimensions following the formulae not for circles, but for spheres. The surface area of a sphere can be calculated using $A=4\pi r^2$. This means that the power available in any signal propagating through three dimensional space drops off at an exponential rate. Twice the nominal designed distance results in less than three percent of the required power. At four times the intended distance, the power has dropped down to an infinitesimal .5% of the original signal. The fact of the matter is that broadcast radio waves are just not practical for use in long range space communications.

To get any value at all out of radio waves in space, the signal must be somehow directed and constricted to a narrow stream of energy which travels in a "line of sight", rather than being allowed to propagate out in all directions. This is how microwaves work, and is essentially how NASA is able to communicate with its long range missions such as the Cassini probe or Hubble. It is absolutely critical that the microwave beam is targeted directly at the spacecraft with very little tolerance for error. Should the craft deviate even slightly from its course, communication is lost until the antenna can be adjusted ever so slightly to retarget it. Using microwave technology, the only signals from deep space that would be able to reach earth, would be signals intentionally aimed directly at earth. These would also be dependent upon the rotation of both the earth and of the source planet, since a signal could only travel in a very straight line between the transmitter on one spinning planet and its intended destination on another spinning planet. Unless both source and destination are located at the respective planet's north or south poles, and those poles just happen to line up, the window of opportunity for such a line is miniscule and is fragmented into bursts where the "daylight" sides of both planets happen to coincide. In other words, for such a signal to reach us, the sender would have had to use very complex algorithms, similar to what NASA currently uses, and would have to know very specific details about the astronomical attributes of Earth, including our exact three dimensional location in the universe at any given time, our rotational direction and frequency, as well as the composition of our atmosphere (to account for and correct for tropospheric scatter).

If they already know this much about Earth, they would have had to have performed advanced remote scanning, the level of which we are only now beginning to achieve (including mass spectrometry chemical atmospheric and surface analysis, the ability to detect compounds such as water or oxygen, and very likely, the ability to predict the existence of life), or they would have had to come here to take such measurements themselves. If they know all these details, they probably know a lot more, and may have already come to the conclusion that no further microwave communication is readily beneficial.

There's another reason why microwaves would not be a good choice. Lasers. Lasers function quite similarly to microwaves in that they are both electromagnetic signals that are artificially created and forced to travel in tight beams. Lasers have all the same drawbacks as microwaves for long distance interplanetary communications. But lasers have one huge advantage. They are a lot easier to detect. Lasers are easily detectable by the human and non human eye. No additional receiving equipment is required at the receiving end. A blinking laser would be readily noticeable to anyone looking into space. The blinking of a laser could be used to encode messages with a rudimentary system such as morse code, or even just a simple pattern of numbers sent as flashes. Our far-off would-be communicators would of course know this. Why in the world would they devise some complicated radio based system when a flashing laser is all that is needed? Lasers work quite well for distracting and entertaining house cats and they would work just as well on humans.

It seems clear to me that no radio waves are being directed to us. No one is trying to talk to us. What's less clear is whether that is because they find us uninteresting, or if they simply are not allowed to talk to us. Ignoring all of the hard science, if we can take anything from mythology it is that steps appear to have been taken to avoid contact, or at least certain types of contact, and at least by certain groups.

The Prime Directive

When Gene Roddenberry cooked up the Prime Directive, he seems to have based it (whether consciously or not) on long standing traditions. The watchers had a job to do that did not involve sexual relations with the locals (Captain Kirk notwithstanding). Contravention of certain cultural contact prohibitions invariably resulted in disagreements, betrayals, punishments and mutinies among the watchers, and sometimes among their human subjects as well.

Policies of noninterference apply beyond relational intimacy. Another very common theme is that of the transfer of technologies from the extraterrestrials to the earthlings. It appears in the myths of the Aztec, the Han, the Hebrews, and the Greeks. Again, this action is generally presented as a contravention of some rule or societal expectation on the part of the extraterrestrials. Sometimes trickery or theft is involved. This crops up repeatedly in the Sumerian mythology. The specific technologies mentioned are often symbolized by abstractions such as fire, or some type of magical object (often a fruit or other food or drink), however this is not always the case. Hebrew accounts included specific skills such as metalsmithing, telecommunications, and the fabrication of as yet undecipherable technologies. Sumerian legends include irrigation, computerized astronomical navigation and advanced bioengineering. Ancient Chinese history speaks of several instances of humans manufacturing flying machines based on 'divine' plans. Is it really possible that all of these stories are based on factual events? Have our visitors tried to teach us, using dumbed down versions of their own technologies? Have we failed to learn time and time again? Is there still much highly advanced tech of the ancient visitors left to discover? Recent meteorite discoveries may point to a clue.

The oldest meteorites discovered on earth are older than the planets. In fact, they have been found to contain small amounts of materials that are older than the Solar system itself.[112] Tiny traces and dusty remnants of ancient supernovae streamed through the birthing clouds of our system around 4.6 Billion years ago, becoming caught in the newly condensing material clumps orbiting baby Sol.

_____

[112] https://en.wikipedia.org/wiki/Presolar_grains

What is truly bizarre about these "presolar grains" is their exotic chemistry which is completely alien to the Solar System. Apart from their unique isotopic ratios by which scientists can determine the type of predecessor star or nebula, they also contain a surprising amount and number of materials whose incredible chemical and physical structures do not occur naturally in our solar system[113]. Such structures include crystalline matrices of silicon carbide[114], single-atom-thick graphene films, and nanodiamonds[115].

Silicon carbide can be manufactured in the lab. It is a very hard ceramic that is used to manufacture automotive brakes and bulletproof vests. It is also used in high voltage semiconductor electronics applications. However, the silicon carbide crystals occurring within presolar grains are several orders of magnitude more pure and structurally stable than any silicon carbide ever grown in the lab.[116]

Nanodiamonds are a relative newcomer to modern material sciences, with active investigation underway for potential application in a wide range of biomedical, electronic, and quantum engineering applications.[117]

---

[113] https://en.wikipedia.org/wiki/Cosmochemistry

[114] https://en.wikipedia.org/wiki/Silicon_carbide

[115] https://en.wikipedia.org/wiki/Nanodiamond

[116] Daulton, T.; Bernatowicz, T. J.; Lewis, R. S.; Messenger, S.; Stadermann, F. J.; Amari, S. (June 2002). "Polytype distribution in circumstellar silicon carbide". Science. 296 (5574): 1852–1855. Bibcode:2002Sci...296.1852D. doi:10.1126/science.1071136. PMID 12052956. S2CID 208322.

[117] Mochalin, V. N.; Shenderova, O.; Ho, D.; Gogotsi, Y. (2011). "The properties and applications of nanodiamonds". Nature Nanotechnology. 7 (1): 11–23. doi:10.1038/nnano.2011.209. PMID 22179567.

By far, the most exotic of these materials is graphene[118]. Graphene is composed of a single layer of carbon atoms, so thin it is sometimes called "two-dimensional". This astounding arrangement of atoms yields a unique mix of attributes including conductivity, flexibility, and transparency; making it highly suitable for a staggeringly long list of potential applications currently under development. It is highly sought after as a lightweight and very strong composite material, and for a range of uses in fields as diverse as electronics, bioengineering, filtration, and ecology.

There are at least twenty-four different ways that graphene can be produced in the lab — all of them very difficult, complicated, or plagued with very low yield rates. In other words, we can make it, but we are really not good at making it. Further development of these methods is being actively investigated.

This makes its appearance within meteorites all the more strange. The presolar grains are very old and from very far away. Neither of these facts truly explain how they ended up containing graphene, a very expensive material with very novel and desirable technological properties for which we have yet to come up with a satisfactory fabrication process. Did someone with more advanced technology than we currently have make this graphene and its associated nanodiamonds and silicon carbide? If so, whatever their end product was has since been pulverized to dust leaving only microscopic shreds of their construction materials. All of this would have had to have

---

[118] https://en.wikipedia.org/wiki/Graphene

occurred billions of years ago and light years away from here. Perhaps it was long long ago, in a galaxy far far away.

If the aliens really are watching, and if they are governed by some galactic prohibition against interference with the human race, then why do we see them so often? On the other hand, if they are under no such prohibition, why don't they make contact in a clear and unambiguous way? Wouldn't you think that an advanced civilization with advanced technology capable of interstellar travel, would be civilized enough to be both law abiding and neighborly? Perhaps not. Or perhaps that is the case for the majority of the members of their civilization, but there are some individuals who like to live a little bit on the wild side.

Considering the aerobatic maneuvers often witnessed, one might conclude that these individuals are showing off or just out for a joyride. Are the aliens who come here no more than little green redneck tourists? Are they just getting drunk and rolling coal and driving like maniacs? That might explain the numerous flying saucer crashes that have turned up over the years. I honestly can't come up with a better explanation for someone to drive all the way across the galaxy, navigating between billions of stars, only to drive straight into the side of a hill in New Mexico. It is entirely possible that the only extraterrestrials we have encountered so far are the renegades and rebels — the free spirits who won't be caged. A lot of the alleged activity associated with UFO encounters is nothing short of bizarre. From cattle mutilations to crop circles and anal probes, one has to wonder what kind of crazy stunts these are.

On the other hand, there are enough examples to counter this argument. Many of the ancient stories seem to involve more intentionality on the part of the aliens. There are many records of extraterrestrial contact with human governments, whether they be kings of old, or current world leaders. Jimmy Carter and Ronald Reagan both witnessed UFOs in flight. So did many kings throughout the historical and ancient records, as discussed previously.

Much of our technology, spirituality, and religion may be based on information given to our leaders by extraterrestrial entities. It is even possible that our very biology can be attributed to them.

Us From Them

For several decades, Professor Chandra Wickramasinghe, a distinguished astrobiologist, has been one of the most vocal proponents of the theory of ongoing panspermia. According to this theory, Earth has been, and continues to be bombarded by ancient microscopic life forms drifting through space. It was these life forms that originally seeded Earth as the first life, and began the process of further evolution in and of our biosphere. The mythological record certainly backs this idea up, as does our own biology.

As for the biology, we humans are organisms made of cells, as you no doubt were taught in school. However, our cells themselves are actually made of other organisms. That's the part they don't teach in school. Well actually, they *do* teach it in school, but not until university level biology programs. According to this theory, a person is actually not "an organism" at all, but rather a complex conglomeration of communities of symbionts derived from simple, single celled prokaryotes such as bacteria. These are not quite the same as the free-living bacteria we all know and love, because they have entered into long standing symbiotic relationships which started billions of years ago when our own ancestors were the earliest eukaryotes. These early eukaryotes formed when one prokaryote swallowed another, but forgot to digest it completely. Prokaryote 'A' swallowed 'B', but rather than breaking it down for nutritional reasons, it instead opted to leave it partly intact and hijack its functionality as its own.

A wonderful example of this process is the fact that all "higher life forms" on earth are either plant or animal.[119] The primary difference between these kingdoms is the mechanism by which the cells convert chemicals to energy in a process called metabolism. Animals use oxygen to fuel their cells, while plants use carbon dioxide. This is because of the two different types of prokaryotes that our forebears ingested and enslaved. Back in the day, bacteria were trying out lots of different chemical reaction chains to create energy. Two of the most successful bacterial processes were photosynthesis which uses sunlight and carbon dioxide, and oxygenation which uses oxygen to "burn" other chemicals in exothermic reactions.

The bacteria who used photosynthesis were the ancestors of what eventually became known as chloroplasts inside the cells of plants, while the oxygenators became the mitochondria inside the cells of animals. Both chloroplasts and mitochondria were taken hostage by other bacteria whose own metabolizing processes were inferior.

---

[119] This section is greatly oversimplified. An accurate accounting of the processes involved is well beyond the scope of this book.

The relationship proved beneficial to both parties in that the chloroplasts and mitochondria were given a cozy little home where all their needs were taken care of. All they had to do was work full time for their host, who guaranteed them protection, an ample supply of fuel, and an eternal lineage for their offspring. Thus was the mitochondria domesticated by our very early ancestors. The mitochondria is not the only simple organism to live in each of our cells, there are many others, which we now refer to as organelles. Each of these was once a separate species, which became domesticated and entered a complex relationship of reciprocal exchanges. This is how the complex eukaryotic organisms formed.

This may sound like a wild idea lacking any evidence or justification but even though it seems to come out of left field for the layman, in fact this theory, now known as "symbiogenesis"[120] or "endosymbiotic theory" is widely accepted by nearly all academic and professional biologists.[121] The greatest piece of undisputed evidence for this theory is the simple fact that mitochondria and other types of organelles actually have their own, completely non-human DNA. These are non-human entities living inside our cells. It is worth pointing out once again the fact that Francis Crick, co-discoverer of DNA, considered these early organelle ancestors, with their already complex DNA to have originated somewhere other than earth.

---

[120] https://en.wikipedia.org/wiki/Symbiogenesis

[121] Cornish-Bowden, Athel (7 December 2017). "Lynn Margulis and the origin of the eukaryotes". Journal of Theoretical Biology. The origin of mitosing cells: 50th anniversary of a classic paper by Lynn Sagan (Margulis). 434: 1. Bibcode:2017JThBi.434....1C. doi:10.1016/j.jtbi.2017.09.027. PMID 28992902.

The DNA evidence is truly a smoking gun, but there is even more astounding evidence that shows not only are these organelles separate creatures, but they also have their own lifecycle completely separate from our own. The organelles within our cells reproduce independently from our own reproduction cycle and even independently from the reproduction of our individual cells. Inside any given cell are organelles dividing, multiplying and spreading their own genes within the environment that is their cellular home. Entire colonies of organelles are thriving within you. These organic entities are not from earth. You are literally a multispecies space colony. Not only that, but the colonies themselves are not primitive, but rather should be thought of as a civilization of intelligent beings.

Bacteria are able to monitor their environment, process information,[122] and intelligently make decisions.[123] Bacterial cells can communicate and alter group behavior with one another. This behavior is recognized as a manifestation of intelligence and self-awareness, and allows entire bacterial networks to recognize and adjust to particular environments collectively. In other words, there is intelligence at work in both the individual bacteria and at the level of the community.[124] It could be argued that at a collective level, bacteria show more wisdom than humans,

---

[122] Pinto, D., Mascher, T. (2016). Bacterial "intelligence": using comparative genomics to unravel the information processing capacity of microbes. Current Genetics. 62: 487-498.

[123] Nakagaki, T., Yamada, H. (2000). Maze-solving by an amoeboid organism. Nature. 407: 470-471.

[124] Steinert, M. (2014). Pathogen Intelligence. Frontiers in Cellular and Infection Microbiology. 4(8): 1-7.

acting for the betterment of their society as a whole. It makes one wonder, if these microbes came from space and planted colonies on earth, and if they are as intelligent as they appear to be, was their colonization of earth and their highly advanced structural symbiosis planned and done *intentionally*?

Summary

As we near the end of our journey through this investigation, let us review briefly some of the topics discussed and what we have learned.

We saw that regardless of the weak statistical methods used to predict the fact, the universe has been observed, examined, studied, and found to be full of an abundance of potentially hospitable planets and other non-planetary environments that do in fact contain many of the factors that we deem suitable for life as we know it.

We saw that organic compounds and biological traces are plentiful and readily discoverable; and in fact already discovered by us, nearly everywhere we have thought to look, including on the moon and Mars, and the uncountable profusion of asteroids, meteors, and comets.

We saw that these traces occur across our solar system and beyond, extending to other stars and nebulae that predate the formation of our sun.

We saw that by scientific methods we have detected fossilized partial corpses of microscopic lifeforms. We have even discovered evidence of living organisms on Mars.

We saw that this information was willingly suppressed by NASA (or perhaps by someone higher up in government who NASA reports to). This suppressed data and its implications has to this day not been publicly admitted by the authorities.

We saw that we have actually placed living creatures on the lunar surface and left them there. Being Tardigrades, they most likely survive even now, albeit possibly in their "hibernating" state.

We saw that organic material is being constantly redistributed through interplanetary and interstellar space.

We saw that there is a very long history of humans witnessing incoming celestial bodies, and other unidentifiable objects that appear to be visiting earth, and that many of these objects appear to be operated intelligently.

We saw that most of the evidence, data, and implications has been essentially ignored by academia and withheld from the public and the media. In some cases the media has been employed as a tool for humiliating and discrediting those who refuse to look the other way, often to the critical detriment of their scientific careers.

We saw that UFO witnesses describe their experiences in ways that are internally consistent within the body of evidence, exemplifying only a small number of possible craft with minor variation in attributes. This has been the case for all of recorded history, including prehistoric anecdotal evidence in the form of oral history.

Are They There? - Yes.

Are They Here? - Yes.

Are there still a lot of questions about who they are, where they have come from, and why they have come? - Yes.

The Paradox Re-examined

We now can return to the "paradox" itself. As you will recall, we identified seven separate statements that comprise the idea.

1. There is a strong probability of the existence of advanced extraterrestrial life.
2. There is a high level of confidence in our ability to obtain evidence of the existence of advanced extraterrestrial life.
3. There is a lack of conclusive evidence of advanced extraterrestrial life.
4. This supposed lack of evidence appears to disagree with statement #2.
5. This seems strange for some reason.
6. Maybe it's a paradox?
7. Therefore, there probably isn't any advanced extraterrestrial life after all.

In light of what we just summarized, let us examine again each of the above statements.

Statement 1 - There is a strong probability of the existence of advanced extraterrestrial life. Probability has nothing to do with it. They are both "out there" and "already here". We have seen them with our own eyes, our radar, our microscopes, and our satellites, and our Mars Rovers.

Statement 2- There is a high level of confidence in our ability to obtain evidence of the existence of advanced extraterrestrial life. This statement is sadly false; to our shame. We have proven to be dismal failures at recognizing the evidence that is right in front of our faces and our scientific instruments. Thankfully, there is hope that this trend is beginning to change. As more and more evidence becomes available to more and more open minded individuals, corporations, and institutions, it may be only a matter of time before even the most stubborn of skeptics are forced to face the truth of the evidence.

Statement 3 - There is a lack of conclusive evidence of advanced extraterrestrial life. This statement is absolutely false. The abundance of evidence presented in this book and many others, and testified to by millions of UFO witnesses and thousands of scientists around the globe all adds to a very strong case for the existence of extraterrestrial life in general, and technologically advanced intelligent life specifically.

Statement 4 - This supposed lack of evidence appears to disagree with statement #2.

Statement 5 - This seems strange for some reason. Both statements 2 and 3 are false. Whether they are confident or not, the only reason a person may perceive a lack of evidence is directly because said person has an amazingly low ability to obtain relevant evidence. Whether two falsehoods agree with each other is logically irrelevant. They may be false for different reasons.

Statement 6 - Maybe it's a paradox? It seems difficult enough to even define a paradox, so I guess maybe it is? But not in my book. (And *this is* my book)

Statement 7 - Therefore, there probably isn't any advanced extraterrestrial life after all. After all this evidence, I don't even know what else to say. "Aliens Must Not Exist" is wishful thinking. "Where Is Everybody?" is an exercise in intentional continued ignorance.

We Are The Champions

Early on in this book, I brought up the point that there is a tendency among skeptics, and especially scholars to discredit new information that seems difficult to fit into their specialized area of expertise. Because of the necessary choices we must all make, we often get ourselves into positions over ever more narrowing focus. We choose a field and a specialty, and it becomes a point of self-identity. We are what we do. We choose fields we enjoy or find interesting. This is great and necessary for advancement of knowledge. However, this all too easily evolves into "my field is the best", which itself evolves into "I am the best". As a species (yes, I see the irony here), we have certainly done this. We think homo sapiens is the one and only, and the smartest thing that ever was. How mistaken we are.

This line of thinking also comes with a dangerous corollary that "my experiences count, yours don't". This is perhaps why so many scientists simply reject the entire UFO phenomenon. This is why they must also reject the brilliant work of cutting edge scientists who have stepped outside the comfort zone and found conclusions that are difficult to believe, even when they are backed by evidence. Perhaps we can start to buck this trend.

Perhaps we can learn from our extraterrestrial friends, neighbors, and ancestors. Perhaps we can embrace our own alienness. Perhaps we can embrace new information from whatever its source, and like our intelligent internal symbionts, make wise decisions and solve communal problems. Perhaps we already have. Perhaps we have simply forgotten our identity. Perhaps regaining this identity is the key to understanding ourselves and others. Perhaps it is the key to our survival both here on earth, and out into the beyond. Perhaps, only once we have learned this lesson will our neighbors be willing and able to speak openly.

About the Author

Ken Goudsward is a systems analyst with expertise in industrial robotics, software engineering, and data design. He enjoys applying these skills to ancient Hebrew and Sumerian documents, and the UFO phenomenon.

Other Books by Ken Goudsward:
- Before Roswell (with Barbara De Long)
- UFOs In The Bible
- Magic In The Bible
- The Enuma Elish; the Original Text With Brief Commentary
- The Atrahasis Epic

Novels by Ken Goudsward:
- Symphony Of Destruction (Hard Sci-Fi)
- Munchausen By Proxy For Fun And Profit (Crime/Dark Comedy)

www.ingramcontent.com/pod-product-compliance
Lightning Source LLC
Chambersburg PA
CBHW071159200326
41519CB00018B/5285